504-B
1989-01

CURRENT OPTIONS FOR CEREAL IMPROVEMENT

ADVANCES IN AGRICULTURAL BIOTECHNOLOGY

Akazawa T., et al., eds: The New Frontiers in Plant Biochemistry. 1983.
ISBN 90-247-2829-0

Gottschalk W. and Müller H.P., eds: Seed Proteins: Biochemistry, Genetics, Nutritive Value. 1983. ISBN 90-247-2789-8

Marcelle R., Clijsters H. and Van Poucke M., eds: Effects of Stress on Photosynthesis. 1983. ISBN 90-247-2799-5

Veeger C. and Newton W.E., eds: Advances in Nitrogen Fixation Research. 1984. ISBN 90-247-2906-8

Chinoy N.J., ed: The Role of Ascorbic Acid in Growth, Differentiation and Metabolism of Plants. 1984. ISBN 90-247-2908-4

Witcombe J.R. and Erskine W., eds: Genetic Resources and Their Exploitation — Chickpeas, Faba beans and Lentils. 1984. ISBN 90-247-2939-4

Sybesma C., ed: Advances in Photosynthesis Research. Vols. I-IV. 1984.
ISBN 90-247-2946-7

Sironval C., and Brouers M., eds: Protochlorophyllide Reduction and Greening. 1984. ISBN 90-247-2954-8

Fuchs Y., and Chalutz E., eds: Ethylene: Biochemical, Physiological and Applied Aspects. 1984. ISBN 90-247-2984-X

Collins G.B., and Petolino J.G., eds: Applications of Genetic Engineering to Crop Improvement. 1984. ISBN 90-247-3084-8

Chapman G.P., and Tarawali S.A., eds: Systems for Cytogenetic Analysis in *Vicia Faba* L. 1984. ISBN 90-247-3089-9

Hardarson G., and Lie T.A., eds: Breeding Legumes for Enhanced Symbiotic Nitrogen Fixation. 1985. ISBN 90-247-3123-2

Magnien E., and De Nettancourt D., eds: Genetic Engineering of Plants and Microorganisms Important for Agriculture. 1985. ISBN 90-247-3131-3

Schäfer-Menuhr A., ed: *In Vitro* Techniques – Propagation and Long Term Storage. 1985. ISBN 90-247-3186-0

Bright S.W.J., and Jones M.G.K., eds: Cerial Tissue and Cell Culture. 1985.
ISBN 90-247-3190-9

Purohit S.S., ed: Hormonal Regulation of Plant Growth and Development. 1985. ISBN 90-247-3198-4

Fraser R.S.S., ed: Mechanisms of Resistance to Plant Diseases. 1985.
ISBN 90-247-3204-2

Galston A.W., and Smith T.A., eds: Polyamines in Plants. 1985.
ISBN 90-247-3245-X

Marcelle R., Clijsters H., and Van Poucke M., eds: Biological Control of Photosynthesis. 1986. ISBN 90-247-3287-5

Semal J., ed: Somaclonal Variations and Crop Improvement. 1986.
ISBN 90-247-3301-4

Purohit S.S., ed: Hormonal Regulation of Plant Growth and Development, Volume 2. 1987. ISBN 90-247-3435-5

Wolfe M.S., and Limpert E., eds: Integrated Control of Cereal Mildews: Monitoring the Pathogen. 1987. ISBN 90-247-3626-9

O'Gara F., Manian S., and Drevon, J.J., eds: Physiological Limitations and the Genetic Improvement of Symbiotic Nitrogen Fixation. 1988.
ISBN 90-247-3692-7

Maluszynski M., ed: Current Options for Cereal Improvement. 1989.
ISBN 0-7923-0064-5

Marshall, G., ed: FLAX: Breeding and Utilisation. 1989. ISBN 0-7923-0065-3

Current Options for Cereal Improvement

Doubled Haploids, Mutants and Heterosis

Proceedings of the First FAO/IAEA Research Co-ordination Meeting on "Use of Induced Mutations in Connection with Haploids and Heterosis in Cereals", 8–12 December 1986, Guelph, Canada

edited by

M. MALUSZYNSKI

Joint/FAO/IAEA Division, Plant Breeding and Genetics Section, Vienna, Austria;
Head of the Department of Genetics, Silesian University, Katowice, Poland

KLUWER ACADEMIC PUBLISHERS

DORDRECHT / BOSTON / LONDON

Library of Congress Cataloging in Publication Data

FAO/IAEA Research Co-ordination Meeting on "Use of Induced Mutations
   in Connection with Haploids and Heterosis in Cereals" (1st : 1986 :
   Guelph, Ont.)
     Current options for cereal improvement : doubled haploids,
   mutants, and heterosis : proceedings of the First FAO/IAEA Research
   Co-ordination Meeting on "Use of Induced Mutations in Connection
   with Haploids and Heterosis in Cereals", 8-12 December 1986, Guelph,
   Canada / edited by M. Maluszynski.
        p.   cm. -- (Advances in agricultural biotechnology)
     Includes bibliographies.
     ISBN 0-7923-0064-5 (U.S.)
     1. Grain--Breeding--Congresses. 2. Grain--Genetics--Congresses.
   3. Plant mutation breeding--Congresses.   I. Maluszynski, M., 1941-
   . II. Title. III. Series.
   SB189.5.F36 1986
   633.1'042--dc19   88-31487

ISBN 0-7923-0064-5

Published by Kluwer Academic Publishers,
P.O. Box 17, 3300 AA Dordrecht, The Netherlands.

Kluwer Academic Publishers incorporates
the publishing programmes of
D. Reidel, Martinus Nijhoff, Dr W. Junk and MTP Press.

Sold and distributed in the U.S.A. and Canada
by Kluwer Academic Publishers,
101 Philip Drive, Norwell, MA 02061, U.S.A.

In all other countries, sold and distributed
by Kluwer Academic Publishers Group,
P.O. Box 322, 3300 AH Dordrecht, The Netherlands.

EDITORIAL

The papers are arranged in alphabetical order. Typing instructions for camera-ready manuscripts were given to authors. The views expressed, nomenclature, and way of presentation remain the responsibility of the authors concerned, who are also responsible for any reproduction of copyright material. In certain cases only a short communication is presented.

The excellent assistance of Mrs. Kathleen Weindl, Joint FAO/IAEA Division, Plant Breeding and Genetics Section, in making corrections in English and retyping some papers, is highly appreciated.

First FAO/IAEA Research Co-ordination Meeting on
"Use of Induced Mutations in Connection with Haploids
and Heterosis in Cereals"

8-12 December 1986
Guelph, Canada

BJÖRN SIGURBJÖRNSSON
Director, Joint FAO/IAEA Division

FOREWORD

The use of induced mutations in plant breeding is now
accepted as a useful means of adding valuable attributes to
a variety. Crop varieties with induced mutants in their
background now number over one thousand. Mutation breeding,
like the backcross or the use of hybrid vigor has thus been
added to the standard tools used by plant breeders. As new
crop varieties become more "thoroughbred" and the demands
for superior performance of crops become more exacting, it
is urgent to explore more sophisticated ways of generating
and manipulating germ plasm. While we wait for the miracles
of "genetic engineering" technqiues to revolutionize plant
breeding, we must explore the potential of known
technologies and combinations thereof. The Co-ordinated
Research Programme organized by FAO and IAEA which is
launched by this Research Co-ordination Meeting explores the
utilization of several modern approaches: induced
mutagenesis, haploidy, heterosis, tissue culture, male
sterility and wide crosses and especially the use of induced
mutagenesis in combination with the other techniques.

As will be seen in the papers significant rsults have
already been achieved and the prospects for successful
harnessing of useful combinations of these techniques for
cereal improvement are bright indeed.

The sponsoring Organizations are interested in
promoting and assisting further research and development in
these and other advanced biotechnologies for improving the
productivity and quality of crops.

It will be interesting to see how far we have
progressed after completing five years of research by the
participating scientists in this Co-ordinated Research
Programme. The results will be published as proceedings of
the third and last Research Co-ordination Meeting.

CONTENTS

QUANTIFYING GAMETOCLONAL VARIATION IN WHEAT DOUBLED HAPLOIDS

P.S. BAENZIGER, C.J. PETERSON, M.R. MORRIS, P.J. MATTERN

Department of Agronomy and Agricultural
Research Service
USDA - University of Nebraska
Keim Hall
Lincoln, Nebraska, USA

M. Maluszynski (ed.), Current Options for Cereal Improvement, 1–9.
© *1989 by Kluwer Academic Publishers.*

Abstract

Doubled haploid breeding methods are being widely used in some crops and at least two wheat cultivars have been developed using doubled haploid techniques involving anther culture. However, many questions relating to the efficacy of wheat doubled haploid breeding methods remain unanswered. Some of the unanswered questions include do the doubled haploids derived using current anther culture techniques accurately represent the gametic array, what is the level of chromosomal rearrangement in doubled haploids, and are truly homozygous, homogeneous doubled haploid lines as environmentally stable as more heterogeneous cultivars developed using conventional breeding techniques. These questions will need to be answered before anther culture and doubled haploidy can be successfully incorporated into a breeding program. This paper provides possible experimental approaches to answering these questions.

Introduction:

Since the initial discoveries of Guha and Meheshwari (1964) and Nitsch and Nitsch (1969) that haploid plants could be obtained from cultured anthers, doubled-haploid theory and methodology have advanced rapidly (reviewed by Collins and Genovesi, 1982; Maheshwari et al., 1982; Baenziger et al., 1984; Choo et al., 1985; Herberle-Bors, 1985). Wheat anther culture methods for obtaining doubled haploids, while not optimal for efficient breeding use, are becoming routine (Deaton et al., 1986; deBuyser et al., 1985; Henry and deBuyser, 1985; Lazar et al., 1984; Marsolais et al., 1984). Recent research efforts in doubled haploids have culminated in the successful development and release of doubled haploid cultivars in France ('Florin') and China ('Jinghua No. 1').

Despite successes and advancement of anther culture and doubled haploid methodology, there is a distinct lack of information concerning critical parameters necessary for applying doubled haploid technology in a predictable and routine manner in plant breeding. A breeder first must know if anther culture derived doubled haploids accurately represent the gametic array of parental materials. In tobacco, aberrant genetic ratios for segregation of qualitative disease resistance traits were found among doubled haploid derived lines (Burk et al., 1979). In rice, normal gene segregation in doubled haploids was shown and used to determine gene linkage relationships (Chung-Morg et al., 1982). Segregation of a qualitative trait, glume color, and a quantitative trait, plant height, in wheat were shown to be normal in doubled haploids derived from simple single cross F_1's (Hu et al., 1979). Henry and de Buyser (1985) believed that genes influencing regeneration of haploid wheat plants may be linked to the 1B/1R translocation, finding a higher than expected frequency of the trans-location in regenerated plants. Mascia (personal communication, 1986) found an increased frequency of short plants, possibly semi-dwarfs and full dwarfs, in doubled haploids derived by anther culture of an F_1 from a tall by semi-dwarf cross and an F_1 from semi-dwarf by full dwarf cross, respectively. The ability to generate doubled haploids through anther culture has been shown to be genotype dependent (Lazar et al., 1984). This may restrict the expected variability of genotypes obtained through anther culture, especially in materials heterozygous for genes affecting cultur-ability and regenerability.

If anther culture derived doubled haploids represent a biased sample of the gametic array, the nature of the bias must be identified and quan-tified before successful breeding strategies can be developed and implemen-ted.

It is well documented that wheat anther culture can induce chromosome loss and breakage (de Buyser et al., 1985; Larkin and Scowcroft, 1983; Kudirka et al., 1983, 1986; Hu et al., 1978). Gross chromosomal changes are readily observed and evident in weak and poorly developed regenerated plants. However, no study to date has determined the frequency of more subtle chromosome rearrangements such as translocations and inversions. Such chromosomal rearrangements at the haploid levels become homozygous and phenotypically indistinguishable upon chromosome doubling. Hexaploid wheat tolerates most translocations because of a high degree of gene duplication among genomes, hence partial sterility is not expected. In other crops, especially diploids, a significant degree of sterility could result. Rearrangements appear in the form of meiotic abnormalities when crossed to other wheats resulting in reduced recombination, gene rearrangements, and aberrant segregation of alleles for genes in the affected chromosomes. If chromosomal rearrangements are found to occur in high frequency, this would hinder the use of doubled haploids as parents for development of improved varieties and restrict use of doubled haploid techniques in recurrent selection programs.

In a conventional wheat breeding program using a modified bulk breeding procedure, the amount of additive genetic variation ($\sigma_a{}^2$) expressed among lines for evaluation and selection depends upon the genera-tion in which the plant selections were made from within a bulk population. The additive genetic variance expressed within a bulk population increases with each generation of selfing, from 1 $\sigma_a{}^2$ as defined among heterozygous F_2 plants, to 2 $\sigma_a{}^2$ among completely homozygous plants. In the Nebraska breeding program, similar to many other programs, plants are selected from within F_3 bulks on the basis of 1.5 $\sigma_a{}^2$ expressed among plants. With successive generations of selfing, the additive variance expressed within F_3 derived lines increases, up to a maximum of 0.5 $\sigma_a{}^2$. This heterogeneity expressed within lines is believed to provide an important population buffering effect contributing to yield stability and wide adaptation. Using doubled haploids, completely homozygous plants would be produced by anther culture of F_1 plants. These derived plants would express all of the poten-tial additive genetic variance (2 $\sigma_a{}^2$) among lines. The concealing effects of dominance variance is eliminated due to lack of heterozygotes. Selec-tion efficiency is expected to increase with higher levels of additive genetic variance expressed among lines.

While the theory is well developed the supporting empirical data is lacking. The supporting data on the Chinese releases has been fragmentary.

The supporting data on the French release (Henry, personal communication, deBuyser et al., 1985) is surprising in the only 41 doubled haploids from the cross that produced the cultivar were field tested. Confirmation of the Chinese and French successes is necessary to further document the potential of doubled haploidy for cultivar development. As described previously, many U. S. wheat breeding programs routinely select heads in F_3 bulks and advance the F_3-derived lines without further selection to variety release. The commonly held belief is that the resultant heterogeneity is beneficial in providing the environmental stability necessary for the broad adaptation commonly found in the Nebraska-developed wheats. The need for heterogeneity has also prompted research in blends of cultivars and in multilines (generally developed for pest resistance, but also for some physiological traits; Frey, 1982; Wolfe and Barrett, 1980). If heterogeneity is necessary for environmental stability, the homogeneous doubled haploid lines will have more limited adaptation than the F3-derived lines. The only research on the environmental stability of doubled haploids found no differences between spring barley doubled haploids derived by interspecific hybridization (Hordeum bulbosum method, Kasha and Kao, 1970) and conventionally derived check cultivars in Ontario environments (Reinbergs et al., 1978). Hence homogeneity was not detrimental for adaptation in that particular study. Similar data for wheat, for doubled haploids derived by anther culture, and for fall or winter sown crops are lacking. Environmental stability and breadth of adaptation are critical parameters for superior cultivar development, especially in highly variable environments.

Some Unanswered Questions Affecting Doubled Haploid Breeding:

The purpose of this section is to identify some of the research areas that need to be explored in the evaluation of the application of modern doubled haploid (anther culture) technology for use in advancing conventional wheat breeding methodology and increasing selection efficiency. The research should address the following questions:

1. Do doubled haploids regenerated from anther culture represent an unbiased sample of the gametic array?
2. What is the frequency of anther culture induced chromosomal rearrangements in doubled haploid lines?
3. Are completely homogeneous, inbred doubled haploid lines derived from anther culture as stable over wide ranges in environments as conventional, more heterogeneous cultivars?

The doubled haploid lines developed to answer the above questions will need to be compared to their equivalent conventionally derived lines by genetic and cytogenetic analyses, and field testing in diverse environments. Information obtained in answering these questions also will provide better understanding of tissue culture-induced variation, cultivar-environment interactions, and cultivar stability. A comprehensive approach will provide the necessary information to determine the true potential and effective approaches for incorporation of doubled haploid technologies into plant breeding programs.

Possible Experiments to Resolve these Questions:

1. Determine if doubled haploids derived from anther culture are an unbiased sample of the gametic array.

   To answer this question, it must be recognized that a biased gametic array can be caused by either or both tissue culture induced variation (somaclonal or gametoclonal variation) or by having some genotypes be more competitive in the callus initiation and regeneration stages (this would be an example of microspore selection). In addition, one should measure the gametic array for both qualitative and quantitative traits. There are many experiments that can be done to resolve this question. The first experiment would be to make a cross between two parental lines that differ for both qualitative and quantitative traits. The resultant F_1 plant would be used for anther culture to develop doubled haploid plants (a minimum of 100 plants is suggested). In addition, the F_1 plant would be used to develop an F_2 population which would be advanced by single seed descent to develop random homozygous plants (probably stopping in the F_6 or F_7 generation, again a minimum of 100 plants is suggested). The doubled haploid lines and the single seed descent lines would be grown in diverse environments and compared for segregation of qualitative and quantitative traits (done using phenotypic and genotypic variances and frequencies). If the variances and frequencies are similar between the doubled haploid and single seed descent lines, then one can assume that the gametic array has been accurately represented using anther culture.

   If there is a difference between the variances and frequencies between the doubled haploid and single seed descent lines, then one can assume that one or more of the following has occurred: 1. the anther culture techniques induce variation, 2. the anther culture techniques preferentially select some microspores for their ability in callus initiation and regeneration, 3. the traits being studied are linked (on a theoretical basis, the only difference between single seed descent and doubled haploid breeding methods is that the latter provides less opportunity for recombination), or 4. the single seed descent system may be preferentially selecting some genotypes. In order to determine if the anther culture techniques induce variation, doubled haploids should be obtained from a homozygous line. If the derived doubled haploid lines and the homozygous line are identical then the doubled haploid techniques do not induce variation. If the doubled haploid lines and the homozygous line are different, then the doubled haploid techniques induce variation. It is important to remember that most cultivars are somewhat heterogeneous, hence should not be considered as a homozygous line. If doubled haploids are derived from random plants within a cultivar then they will need to be compared to a random sample of plants or their selfed progeny from the cultivar. Also, it should be recognized that many tissue culture techniques have carry-over effects on the regenerated plants. Hence only progeny of regenerated plants (removed from the carry-over effects of tissue culture) should be used in these comparisons.

   If the anther culture techniques are preferentially selecting microspores for their ability in callus initiation and regeneration, then doubled haploids derived from an F_1 will be skewed for some genotypes. In addition, all of the doubled haploid genotypes should be within the range of single seed descent genotypes. Examples of preferential selection of some microspores include aberrant genetic ratios for qualitative traits and skewed population frequencies for quantitative traits when compared to the single seed descent population. If the double haploid population, when compared to single seed descent population, exhibits transgressive segregation, then it must have additional variation which has been induced by the anther culture techniques.

Differences between the doubled haploid population and the single seed descent population due to linkage effects can be identified by comparing known linkage relationships. The single seed descent lines should have more opportunity for recombination and should have a higher proportion of recombinant genotypes.

Single seed descent breeding methods are intended to provide rapid inbreeding and an unbiased sample of the possible genotypes from a cross. However, some of the techniques used to speed generation advance can exert selection pressure on the population. An example of unwitting selection can be illustrated by considering an F_2 population that is grown in very small pots to conserve space. Genotypes that cannot tolerate crowded conditions would be preferentially lost. Space constraints would not be as great a factor in developing the doubled haploid population because only one generation of increase is needed, hence no or less selection may have occurred in the derivation of doubled haploid lines. In order to determine if single seed descent provides an unbiased representation of the gametic array, the single seed descent lines will have to be compared to a truly unselected population. No differences between the the single seed descent lines and the unselected population would indicate that the single seed descent procedure provides an unbiased representation of the possible genotypes for a cross.

2. Determine the frequency of anther culture induced chromosomal re-arrangements in anther culture derived doubled haploid lines.

To answer this question, it must be recognized that chromosomal rearrangements at the haploid level will become homozygous (hence phenotypically indistinguishable) in the doubled haploid. Contrary to what has been done in wheat somatic culture systems, the study of meiotic configurations in doubled haploids will not identify rearrangements . The best experiment would be to derive doubled haploid lines from a homozygous line with known regular meiotic pairing. The resultant doubled haploid line would be crossed to the parent line or its progeny and the meiotic configurations of the F_1 will be studied. The parent line is the control by which the doubled haploids are measured. If doubled haploids are derived from random plants within a cultivar which may be heterogeneous, the experiment should be modified. The doubled haploids would need to be crossed to a random set or to a unique plant within the cultivar. In addition, a random set of plants from the cultivar should be intercrossed or crossed to a unique plant within the cultivar. The latter set of crosses is the control for the doubled haploid crosses. Both the control and doubled haploid crosses would need to have the F_1 plants analyzed at meiosis. If the doubled haploid crosses had more or the same number of chromosome alterations, then the anther culture technique induced or did not induce chromosomal rearrangements, respectively.

3. Determine if homogeneous lines are as environmentally stable as heterogeneous lines.

   To answer this question, the materials previously developed to determine if the gametic array is accurately represented can be used. The
   parents, the doubled haploid lines, and the single seed descent lines
   derived from a cross or from a cultivar should be grown in diverse, but
   representative environments (a minimum of six environments is recommended). The stability of agronomic performance can be determined
   using a number of different methods (for example Eberhart and Russell,
   1966). As the among line additive genetic variance increases and the
   within line additive genetic variance decreases with each generation of
   inbreeding, consideration should be given to deriving the single seed
   descent lines from different generations. For example, some breeding
   programs develop cultivars from F_3 derived lines (at homozygosity, the
   among line additive genetic variance is 1.5 σ_a^2 and the within line
   additive genetic variance is .5 σ_a^2). These derived cultivars are very
   heterogeneous. Other breeding programs require a much higher level of
   homogeneity and derive lines from F_6 generation (at homozygosity, the
   among line additive genetic variance is 1.94 σ_a^2 and the within line
   additive genetic variance is .06 σ_a^2). By comparing the doubled
   haploids to F_3 and F_6 derived lines for stability parameters, one can
   determine if the methods are equivalent for developing stable cultivars. One will also develop good estimates of stability for breeding
   lines developed by the diverse selection schemes commonly used by plant
   breeders.

   It should be recognized that the experiments suggested in this paper
   are one way of answering these questions. There are other ways which
   may be equally good.

Potential limitations:

   As mentioned previously, doubled haploid technology for wheat is
becoming routine. However, significant genotype by anther culture interactions are known to occur and some genotypes can be cultured only at low
frequencies. A critical limitation in this research will be the development large numbers of doubled haploids. This concern can be somewhat
alleviated if the parent lines have been selected so that at least one
parent in each cross is known to be highly responsive for anther culture.
Anther culturability is a heritable trait (Bullock et al., 1982; Lazar et
al., 1984; Henry and deBuyser, 1985; Deaton et al., 1986) so that a high
level of success in generating doubled haploid plants can be expected. A
second limitation is that as new anther culture methods are developed, the
doubled haploids will need to fully evaluated using the procedures
described herein.

8

References:

Baenziger, P. S., D. T. Kudirka, G. W. Schaeffer, and M. D. Lazar. 1984. The significance of doubled haploid variation. In "Gene manipulation in plant improvement." J. P. Gustafson, ed., Plenum Publishing Corp. p. 385-414.

Bullock. W. P., P. S. Baenziger, G. W. Schaeffer, and P. J. Bottino. 1982. Anther culture of wheat F1's and their reciprocal crosses. Theor. Appl. Genet. 62:155-159.

Burk, L. G., J. F. Chaplin, G. V. Gooding, and N. T. Powell. 1979. Quantity production of anther-derived haploids from a multiple disease resistant tobacco hybrid. 1. Frequency of plants with resistance or susceptibility to tobacco mosaic virus (TVM), potato virus Y (PVY), and root knot (RK). Euphytica 28:210-218.

Choo, T. M., E. Reinbergs, and K. J. Kasha. 1985. Use of haploids in breeding barley. Plant Breeding Reviews 3:219-252.

Chung-Morg, C., C. Chi-Chang, L. Mign-Hwa. 1982. Genetic analysis of anther-derived plants of rice. J. Hered. 73:49-52.

Collins, G. B. and A. D. Genovesi. 1982. Anther culture and its application to crop improvement. In "Applications of plant cell and tissue culture to agriculture and industry." D. T. Tomes, B. E. Ellis, P. M. Harvey, K. J. Kasha, and R. L. Peterson, eds. Univ. of Guelph, Guelph, Ontario. p. 1-24.

Corbin, T. C. and E. A. Wernsman. 1985. Comparison of genetic gains from Maternal dihaploid, S. and FS recurrent family selection. Agron. Abstracts p. 50-51.

Deaton, W. R., S. G. Metz, T. A. Armstrong, P. N. Mascia, S. A. Sato, and M. S. Wright. 1986. Genetic analysis of the anther culture response of three spring wheat crosses. Agron Abstr. p. 147.

deBuyser, J., Y. Henry, and G. Taleb. 1985. Wheat androgenesis: cytogenetical analysis and agronomic performance of doubled haploids. A. Pflanzenzuchtg. 95:23-34.

Eberhart, S. A., and W. A. Russell. 1966. Stability parameters for comparing varieties. Crop Sci. 6:36-40.

Frey, K. J. 1982. Multiline breeding. In "Plant improvement and somatic cell genetics. I. K. Vasil, W. R. Scowcroft, and K. J. Frey. Academic Press. New York. p. 44-71.

Griffing, B. 1975. Efficiency changes due to use of doubled haploids in recurrent selection methods. Theor. Appl. Genet. 46:367-386.

Guha, S. and S. C. Maheshwari. 1964. In vitro production of embryos from anthers of Datura. Nature 204:497.

Heberle-Bors, E. 1985. In vitro haploid formation from pollen: a critical review. Theor. Appl. Genet. 71:361-364. Henry, Y. and J. deBuyser. 1985. Effect of the 1B/1R translocation on anther culture ability in wheat (Triticum aestivum L.). Plant Cell Rep. 4:307-310.

Henry, Y. and J. deBuyser. 1985. Effect of the 1B/1R translocation on anther culture ability in wheat (Triticum aestivum L.). Plant Cell Rep. 4:307-310.

Hu, H., T. Hsi, and S. Chia. 1978. Chromosome variation of somatic cells, of pollen calli and plants in wheat (Triticum aestivum L.). Acta Genet. Sin. 5:23-30.

Hu, H., Z. Xi, J. Zhuang, J. Ouyang, J. Zeng, S. Jia, X. Jia, J. Jing, and S. Zhou. 1979. Genetic investigations on pollen-derived plants in wheat (Triticum aestivum). Acta Genet. Sinica 6:322-331.

Kasha, K. J. and K. N. Kao. 1970. High frequency haploid production in barley (Hordeum vulgare L.) Nature 225:874-876.

Kudirka, D. T., G. W. Schaeffer, and P. S. Baenziger. 1983. Cytogenetic characteristics of wheat plants regenerated from anther calli of 'Centurk.' Can. J. Genet. Cytol. 25:513-517.

Kudirka, D. T., G. W. Schaeffer, and P. S. Baenziger. 1986. Wheat: genetic variability through anther culture. Y. P. S. Bajaj, ed. Springer-Verlag, Berlin. Biotechnology in Agriculture and Forestry 2:39-54.

Larkin, P. J. and W. R. Scowcroft. 1983. Somaclonal variation and crop improvement. In "Genetic engineering of plants." T. Kosuge, C. P. Meredith, and A. Hollaender (eds.) Plenum Publishing Corp. p. 289-314.

Lazar, M. D., G. W. Schaeffer, and P. S. Baenziger. 1984. Combining abilities and heritability of callus formation and plantlet regeneration in wheat (Triticum aestivum L.) anther cultures. Theor. Appl. Genet. 68:131-134.

Maheshwari, S. C., A. K. Tyagi, and K. Malhotra. 1980. Haploids from pollen grains - retrospect and prospect. Amer J. Bot. 69:865-879.

Marsolais, A. A., G. Seguin-Swartz, and K. J. Kasha. 1984. The influence of anther culture cold pretreatments and donor plant genotypes on in vitro adrogenesis in wheat (Triticum aestivum L.). Plant Cell Tissue Organ Culture 3:69-79.

Nitsch, J. P. and C. Nitsch. 1969. Haploids plants from pollen grains. Science 163:85-87.

Reinbergs, E., L. S. P. Song, T. M. Choo, and K. J. Kasha. 1978. Yield stability of doubled haploid lines of barley. Can. J. Plant Sci. 58:929-933.

Wolfe, M. S., and J. A. Barrett. 1980. Can we lead the pathogen astray? Plant Dis. 64:148-155.

BREEDING ON A CELLULAR LEVEL, AND RESEARCH ON F_1 HYBRID DEVELOPMENT

Z. BARABAS, Z. KERTESZ, L. PURNHAUSER,
F. SAGI and J.PAUK,

Cereal Research Institute
6701 Szeged, POB 391, Hungary

M. Maluszynski (ed.), Current Options for Cereal Improvement, 11–18.
© *1989 by Kluwer Academic Publishers.*

ABSTRACT

In different somatic cultures initiated from immature embryos or young inflorescences the genotype-dependence for callus induction was almost completely eliminated, though some exceptions were observed. Plant regeneration from any wheat genotype was also achieved.

Haploid cell cultures were established and will be produced from these suspension cultures in the near future. Protoplast isolation from a suspension culture was also successful, but plant regeneration was not obtained so far.

Barley x wheat intergeneric hybrids were produced and from these 22 somaclones were regenerated. Significant morphological differences, such as reduced height and lodging resistance appeared among the BC progenies. Streptomycin resistant potential mutants were isolated from somatic cultures using a special selection system. The test of the putative mutants is under way.

An essentially new, patented technology was worked out for F_1 hybrid seed production in blendings. It also works well in cases of partial male or self-sterility. This system works on cms, gms, si and cha male sterile systems.

*   Research carried out in association with IAEA under Research Contract No. 4462/RB

MATERIAL AND METHODS

Callus cultures

Wheat plants were regenerated from various winter cultivars. Surface-sterilized inflorescences (0,5 - 2,0 mm) were used to obtain callus cultures on the Murashige-Skoog nutrient medium containing different auxins. After 2 - 3 subcultures, regeneration was induced on the same medium at 0.1 mg/l 2,4-D or cytokinins. Regenerated plants were transferred to pots with sterile soil and grown to maturity in the greenhouse. Selection in R_0 and R_1 progenies was made in the greenhouse. Further selections were carried out in the nursery, first by the ear-two-row method and later in advanced plots. Bread wheat 'somaclones' were selected regarding their yield ability or other agronomic characters and tested for baking quality and protein content.

Anther culture

Wheat plants from cross breeding programmes were used for anther culture techniques. Excised anthers contained microspores in the mid-uninucleate stage. The microspore stage was determined for each combination by microscopic observation. Anthers were excised from the florets under sterile conditions and placed into a 0.7% agar medium containing a potato extract (Research Group 301, 1976). For plant regeneration a 190-2 medium (Zhuang and Jia, 1983) supplemented with indole-3-acetic acid (0.5 mg/l) and kinetin (0.5 mg/l) or BAP (0.5 mg/l) was used. Chromosome duplication was carried out by treatment in 0.05% colchicine and 2% DMSO tap water for 5 hours. Plants were raised in greenhouse under optimal conditions. The following progenies of doubled haploids were grown in a wheat nursery.

Suspension and protoplast culture

Wheat cell suspension were initiated from microspore derived embryo-like structures or young, haploid inflorescence in liquid potato induction medium. The maintenance, isolation and culture was carried out according to Thompson et al. (1986).

Barley and wheat hybridization

Hordeum vulgare (GK64) x Triticum aestivum hybrids were produced in the greenhouse from January to March 1985 under temperatures varying between 15 - 25°C, with 16h photoperiod. The hybrid embryos were grown on BM-2 medium (Norstog, 1973).

Resistance to antibiotics

Callus cultures from immature embryos were grown on selective medium containing streptomycin (200 mg/l). For callus induction Murashiege and Skoog medium, supplemented with 1 mg/l 2,4-D, was used. For shoot regeneration 2,4-D concentration was step by step decreased (0.5 up to 0.0 mg/l) and $AgNO_3$ (10 mg/l) was also used as described by Purhauser et al. (1987).

RESULTS AND DISCUSSION

1.   Biotechnological research and breeding

1   Decreasing genotype dependence in wheat callus cultures

Generally our callus cultures were initiated from two main inoculums: immature, approximately 14 day old embryo or young inflorescence (0.8 - 1.2 cm in length). With both, sources somatic cultures can be obtained in practically 100% of the cases. We succeeded to eliminate the previously observed genotype-dependence for callus induction by changing the medium, modifying hormone concentrations and composition, and adding natural extracts to the medium. Among the 63 wheat genotypes studied, belonging to two species, exceptions were found only for varieties "Solaris" (T. durum), "CA 8055" and "GK Istvan" (T. aestivum). Calli can be obtained from these varieties as well, though at a significantly lower rate.

1.2 Plant regeneration
Mixed type regeneration (embryogenic and meristematic) was observed from inflorescence with genotype dependence from 40-80%.

Embryogenic callus induction, maintenance and utilization in effective plant regeneration appeared in literature as an outstanding achievement in the last years. Plants regenerated directly from embryogenic calli are completely identical to the parent material. Neither cytological nor morphological differences have been found in several hundred examined plants. On the basis of our experiments we consider it important to start embryogenic cultures from completely dedifferentiated calli.

1.3. Somaclonal variation
Varieties with a high genetic variability demonstrate higher somaclonal variation after subculturing as well. Homogenic varieties usually show lower somaclonal variation (Galiba and Kertesz, 1985). However, unique SC variants can be obtained from calli which cannot be found among those originating from sexual reproduction, e.g., the stocks of variety Mv 4 have a height of 100-120 cm, while dwarf stocks of 60-65 cm height can be often found among SC progenies. Their gliadin and glutenin subunit pattern did not differ from that of the Mv 4.

An experiment is being carried out to study the real somaclonal variation. Doubled haploids were dedifferentiated in callus culture. During one year regenerants were obtained every four months (4, 8, 12). The SC_1 plants were raised and the SC_2 plants were used for the estimation of somaclonal (biochemical, morphological) variation.

1.4 Establishment of haploid cultures
It seems that positive selection for mutants (sub-culturing, mutagen treatment) will be more efficient on

a haploid than on a diploid level. Mutagenesis increases the genetic variability of the cell population under in vitro conditions. In the first step 270 haploid plants were tested for in vitro manipulation.

Two haploid T. aestivum lines, HD 104, HD 208, which had excellent culture characters (good callus and plant regeneration) were found in this wheat haploid clone population. From these lines long-term (6 month) high-frequency plant regeneration (80%) and somatic embryogenesis in callus culture can be induced. The fact that the HD 208 line is available also on a diploid level makes this work even more complex. Single cell cultures will be produced in the near future.

1.5 Protoplast isolation traits in common wheat

Protoplast isolation was carried out from mesophyll cells of somatic regenerants. The protoplast isolation was successful. Until now, protoplast subculturing has failed. The establishment of a haploid cell culture was attained in a different way as well.

An actively dividing haploid suspension culture was established starting from embryo-like structures of 100 genotypes. This cell culture called "4F" preserved its (albino) plant regeneration capacity even after one year. Protoplast isolation from the 4F culture was successful, but plant regeneration was not obtained as yet.

1.6 Gene transfer in vivo and in vitro between barley and wheat

Since the method of protoplast fusion for cereals has not yet been developed, we try to carry out combined procedures for gene transfer by plant tissue culture methods. Such type of work is being carried out on barley x wheat intergeneric hybrids. Barley lacks the useful gene of lodging resistance, and the tolerance to Fusarium nivale of barley would be helpful in wheat.

A Hordeum chilense x Triticum turgidum hybrid was obtained from Spain (Martin and Sanchez-Monge, 1980). Two cytologically controlled Hungarian barley and bread wheat H. vulgare (GK64) x T. aestivum hybrids were produced.

At present, cytological and biochemical examinations are being carried out. The barley-wheat hybrids were back-crossed with wheat and 15 BC hybrid seeds were obtained from 12,000 pollinated flowers, from which 13 healthy plants are growing. The programme of hybrid back-crossing with barley was unsuccessful.

Twenty-two somaclones were regenerated from Hordeum chilense x Triticum turgidum hybrid. Significant morphological differences were observed, such as reduced height and lodging resistance.

1.7 Resistance to antibiotics

Advanced techniques of tissue culture (e.g. protoplast fusion) cannot be imagined without appropriate selectable markers. Streptomycin have a strong bleaching effect on shoots regenerated from calli. Up to this time we selected 5 putative mutants (which produced green shoots on selective medium) from callus cultures initiated from 20,000 embryos. Tests for streptomycin resistance of their progenies are in progress.

2.   Hybrid varieties in cereal species

Although nearly half a century had passed since the successful introduction of hybrid maize, the practical utilization of the heterosis effect has been limited only to a few species (sorghum, rice). There are a number of limitations (Frankel and Galun, 1977; Frankel, 1983; Lehmann, 1986) but basically 3 problems are to be solved:

- the type of male sterility (genic, cytoplasmic, chemically induced, self sterility)
- the level of heterosis effect
- the cost of hybrid seed production

2.1 Heterosis and F_1 hybrid seed production

It is an established fact that special cross combinations of wheat show significant heterosis in grain yield. The crucial question arises whether this overproduction can compensate for the high costs of F_1 hybrid seed production. This cost in the case of cereal species is of primary importance. The basic problem is that the seed production of wheat or barley is about double that of maize, but the seed rate per unit area is 8-10 times higher than that for maize or sorghum. One of the main limitations of seed production originates from the strip method itself, which works well in maize or sorghum but is less adequate for some autogamous crops, e.g. rice, tomato, etc. In China, rice is planted in alternative male-female rows and supplementary pollination is done by hand (Virmani and Edwards, 1983). Blending of parental lines sometimes would solve this problem, as the flowering parents would stand side by side.

2.2 New method for hybrid seed production using partially male or self-sterile lines

If the blending hybrid seed production has advantages over the striped methods it is worth working out adequate ways to utilize these advantages. A number of theories have been worked out based on plant height, xenia, tolerance to herbicides, seed color markers, etc.

These methods did not prove to be of practical importance for a number of reasons (climatic dependence, partial fertility, environment pollution, difficulties in producing parental lines, homogeneity, female fertility, side effects of cytoplasms, etc.).

In August 1986 a basic new technology for producing hybrid seeds in blendings was patented (Barabas, Z., 1986; Reg. No. OTH 3683/86). Details cannot be revealed at present because of the patent procedure but some of the characters of this method in comparison to others can be given:

- Parents can be planted together in a blending. So planting and harvesting will be very simple. The seed setting on the female will be high.
- It works well even in cases of partial male sterility (cms, gms) or self sterility (si) as well, even in a genic male-sterile system, when 50% or more of the female flowers are fertile, e.g., the F_1 of (ms ms x Ms ms).
- It does not need any post-harvest work
- It can be used in all sexually propagated species, and is applicable in the partly asexual ones, too.

The introduction of this procedure will take only a few years first of all in diploid species. In most cases, it needs breeding and biotechnological work similar to that which is necessary today to improve a new variety in self-pollinated species (barley, rice, etc.). But the investment will be much less than to build up BTT system in barley or XYZ system in wheat.

References

BARABAS, Z. (1986) Patent Reg. No. OTH 3683/86.

FRANKEL, R. (1983) Heterosis. Springer-Verlag, Berlin, Heidelberg, 290.

FRANKEL, R., GALUN, E. (1977) Pollination Mechanisms, Reproduction and Plant Breeding. Springer-Verlag, Berlin, Heidelberg, 281.

GALIBA, G., KERTESZ, Z., SUTKA, J., SAGI, L. (1985) Differences in somaclonal variation in three winter wheat (T. aestivum) varieties. Cereal Res. Comm. 13: 343-350.

LEHMANN, L.C. (1986) Hybrid breeding in small grain cereals: Barley, wheat and rye. G. Olsson, Svalöf 1886-1986. Research and Results in Plant Breeding. Ltd. Jörlag, Stockholm, 291.

MARTIN, A., SANCHEZ-MONGE LAGUNA, E. (1980) Hybrid between Hordeum chilense and Triticum turgidum. Cereal Res. Comm. 8: 349-353.

NORSTOG, K. (1973) New synthetic medium for the culture of premature barley embryos. In vitro, 9: 307-308

PURNHAUSER, L. MEDGYESY< P., CZAKO, M., DIX, P.J. and MARTON, L. (1987) Stimulation of shoot regeneration in Triticum aestivum and Nicotiana plumbaginifolia Viv. tissue cultures using the ethylene inhibitor AgNO$_3$. Plant Cell Rep. 6: 1-4

Research Group 301, (1976) A sharp increase of the frequency of pollen plant induction in wheat with potato medium. Acta Genet. Sin. 3: 25-31.

THOMPSON, J.A., ABDULLAH, R., COCKING, E.C. (1986) Protoplast culture of rice (Oryza sativa L.) using media solidified with agarose. Plant Science, 47: 123-133.

VIRMANI, S.S. and EDWARDS, I.B. (1983) Current status and future prospects for breeding hybrid rice and wheat. Advances in Agronomy, 36: 145-214.

ZHUANG, J.J., JIA, S. (1983) Increasing differentiation frequencies in wheat pollen callus. In: Cell and tissue culture techniques for cereal crop improvement. Sci. Press, Beijing, China, 431

USE OF HIGH PAIRING WHEAT MUTANTS FOR THE TRANSFER OF USEFUL TRAITS FROM ALIEN SPECIES INTO CULTIVATED WHEATS. *

C. CEOLONI, G. DEL SIGNORE, O. BITTI

E.N.E.A., Dip. Agrobiotecnologie
C.R.E. Casaccia
C.P. 2400 - 00100 Rome, A.D., Italy

*   Contribution No. 115 of Dip. Agrobiotecnologie, ENEA, Casaccia, Rome.   Part of this research was conducted under IAEA Research Agreement No. 4463/CF.

M. Maluszynski (ed.), Current Options for Cereal Improvement, 19–30.
© 1989 by Kluwer Academic Publishers.

ABSTRACT
Transfer of useful genes from alien species into cultivated wheats by using wheat mutants that promote homoeologous pairing is a realistic approach. In spite of a number of successful achievements and of the availability of suitable genetic materials, such a research area is however still scarcely exploited.

Progress results are presented here of the transfer to common wheat of a T. longissimum gene conferring resistance to wheat powdery mildew. The ph1b mutation has been the pairing induction system adopted here to promote gene exchange between the alien telocentric bearing the selected gene and each of its homoeologous group-3 wheat chromosomes. The resistant recombinant lines, obtained with a frequency ranging from 5.2 to 8.2% depending on the type of cross combination, are being characterized and progressively brought into chromosomal stability and euploidy Crosses have been started to incorporate mildew resistance, now in the background of the variety Chinese Spring, into that of good yielding Italian wheats.

Preliminary results are also reported on the development of a common wheat ph1 + ph2 double mutant. Such a line has been considered worth synthesizing to definitely ascertain if the observed ph1 ceiling to homoeologous pairing can be raised any further by the contemporary presence of ph2, particularly when very distant genomes or single chromosomes are induced to pair with those of wheat. Whether or not an additive effect is exerted by the two mutations is being verified both in wheat itself and in different wheat-alien species hybrids.

Significance of the overall problem

Increasing crop yield has been man's major concern since the times of domestication of the most important plant species, including of course, cereals such as wheat, corn, barley, etc. Plant breeders of this century have been successful in bringing about spectacular progress in this respect. However, a further rise in the yield ceiling and, even more, in yield stability are certainly desirable. Disease and pest control and improved tolerance to adverse environmental conditions remain the main objectives. Varieties also need to be bred for different socio-economic and agro-ecological situations, being so able to respond both under low and high input conditions. It is hard to imagine significant advances in crop improvement using only strictly conventional approaches.

A parallel consideration to this has to be made, i.e. that the genetic material of several main crops, notably wheat, has already been exploited almost to its full capacity. Restoration and enrichment of the cultivated gene pool can be successfully accomplished by resorting to the vast genetic resources present in the wild relatives of alien species related to the cultivated ones. The wide range of useful genes they contain (e.g. resistance to several diseases, tolerance to environmental constraints, nutritional and technological qualities, etc.) certainly deserves consideration, not only as a scientific curiosity but also for its potential for breeding purposes.

Actually, the possibility of using interspecific and intergeneric gene introgression as a breeding strategy is being taken more and more into account so that even Institutions operating on a world scale, like CIMMYT, have put considerable efforts into it [1]. Indeed, wide crosses, with or without subsequent backcrosses to the cultivated parent, have the potential to provide an array of novel genetic combinations, which can not only satisfy specific selection objectives, but also show up unpredictable characters (e.g. vegetative heterosis, nucleo-cytoplasmic interactions resulting in male sterility, etc.).

Certainly the necessary genetic and cytogenetic knowledge and methodologies to make use of the wild/alien germplasm are not available for all crop species, nor is such germplasm always at disposal in the form of more readily usable material (e.g. alien genomes "fractionized" into addition or substitution lines of single chromosomes into a cultivated background).

In this respect wheat is an amendable species: an extremely wide array of aneuploid lines are available in the background of cultivated types (particularly common wheat, T. aestivum), including the above-mentioned addition and substitution lines. Such lines can be much more efficiently used than the original wheat-complete alien species hybrids in works aimed at transferring only selected characters from the alien source. More than this, different strategies have been successfully attempted which permit a further reduction

- from complete chromosomes or chromosome arms to chromosome segments - of alien chromatin to be introduced into cultivated genotypes (see e.g. 2 and 3).

It thus appears to be a feasible task to provide breeders with lines where the amount of alien material still bearing the desired trait(s) is conveniently limited, with the possibility of the contemporary presence of unwanted genes brought to a minimum.

Chromosome engineering via ph1 wheat mutants

Among the different approaches that can be followed to engineer wheat chromosomes, so to obtain transfer of only selected portions of alien chromatin, the one that ensures the greatest degree of "precision" is induction of pairing and recombination between homoeologous (functionally related) chromosomes of the cultivated and the alien genome(s).

Such a strategy relies on the knowledge of the genetic basis of the prevention of homoeologous pairing, a phenomenon which normally occurs in polyploid wheats and in their hybrids with related species. In fact, since suppression of this type of pairing is mainly due to the Ph1 wheat gene located on chromosome 5B, to induce alien chromosomes to pair with their wheat homoeologues one can either operate with a 5B-lacking background or, better, use the ph1 mutants which have been obtained in T. aestivum [4], T. durum [5] and triticale [6]. Either way, homoeologous recombination products can be recovered which are, as such, likely to be more balanced than transfers derived from random events induced, for instance, through radiations.

Following the pioneer work of Sears, who transferred two different leaf rust resistant genes from two Agropyron elongatum chromosomes to their wheat homoeologous 3D and 7D [2,7,8,9], homoeologous pairing has been induced in the absence of 5B or in the presence of a ph1 mutation in various other wheat-alien chromosome combinations, in several cases involving Agropyron chromosomes carrying disease resistant genes [10,11,12 and Tab. 3]. Several amphidiploids have also been developed between the ph1 mutant of durum wheat and different Aegilops species as well as Secale cereale [6,13]. Besides this, the durum wheat-rye ph1 amphidiploid, which is in practice as a 6X ph1 primary triticale, has been employed in crosses with ph1b common wheat and various B genome lacking amphidiploids, all carrying a D genome, in order to turnover into hexaploid triticale qualitative traits (e.g. genes for bread-making quality) controlled by D genome chromosomes [14].

Progress results are presented here of the recently obtained transfer to common wheat of a T. longissimum (Ae. longissima) dominant gene conferring resistance to powdery mildew [3,15]. Briefly recalling the main steps of the transfer procedure adopted, a common wheat cv. Chinese Spring ditelosomic addition line for the critical alien chromosome arm [16] was used to pollinate Chinese Spring double monosomics, having in monosomic condition chromosome

5B and, in turn, each of the group-3 wheat chromosomes, previously been established to represent the homoeologous counterparts of the selected longissimum chromosomes (3S[1]s) [17]. By susbstituting the normal 5B (Ph1) with one carrying the ph1b mutation, a step which was performed by crossing the F_1's between the above lines and the Chinese Spring ph1b mutant, homoeologous pairing has been promoted, possibly in a preferential way between the alien telo and its unpaired homoeologue (3A,3B or 3D, depending on the initial cross) [3]. The frequency of recombinant resistant plants recovered after cross with the euploid averaged from 5.2 to 8.2% (Tables 1 and 3). This infers that the actual pairing frequency between the critical homoeologues (four times the gametic recombination frequency, i.e. about 20 to 30%) largely exceeded the estimates based on meiotic M_I records, which as in similar experiments [2,10,11], were made rather difficult by chromosome clumping.

Of the 12 resistant recombinants so far isolated out of 169 plants tested, 9 appear to be of the "proximal" type (Table 1), i.e. having only complete chromosomes (the resistance was originally located on the alien telocentric). Of these, only one shows "rachis fragility", an undesired character controlled by a gene(s) located on the same alien arm.

When in competition with normal gametes from euploid Chinese Spring, male transmission of the transfer chromosomes tested so far (Table 2) does not seem to be dramatically reduced, though to some extent it is present in all of them. M_1 pairing configurations of the heterozygous recombinants in two subsequent BC generations to be the euploid show a progressive increase in regularity: some of them, indeed, now exhibit 21 bivalents, all of the ring type in some cells (Table 2), thus indicating the occurrence of normal pairing between the transfer chromosome and its wheat counterpart.

Tests for further characterization of the recombinant lines, including the identification of the wheat chromosome involved in each of them, are under way. Crosses have also been started to introduce mildew resistance into good yielding Italian common wheat varieties.

Future developments

The experiments carried out so far indicate that the levels of wheat-alien homoeologous chromosome pairing and, consequently, the recombination frequencies obtainable by using ph1 mutations, although always usable, in some cases appear to be remarkable, whereas in some others are relatively low (Table 3). This may depend on the particular homoeologous group concerned and/or on the cytogenetic affinity of the alien genomes with respect to those of wheat. Particularly low, for instance, is pairing and recombination between wheat and rye chromosomes, even under ph1 promoting conditions [18,19]. In such cases, where particular "reluctance" to pairing is observed, it may be convenient to resort to a non-genetic strategy (e.g. pollen or seed irradiation). However, further manipulations of the complex genetic system controlling pairing in wheat and wheat hybrids are possible and their potential needs to be ascertained.

Ph1, indeed, is the major but not the only suppressor of homoeologous pairing. Of the additional ones Ph2, located on common wheat chromosome arm 3DS, is the most effective, having about half the strength of Ph1. Since ph2 mutants are also available [20,21], a work plan has been conceived aimed at combining the two independent pairing mutations in a single genotype. This would allow verification of whether they can operate in an additive way and, thus, if the observed ph1 ceiling for homoeologous pairing can be raised by ph2, mainly when chromosomes of relatively distant genomes are brought into pairing. Even for more closely related wheat-alien combinations, the size of the populations to be screened for the recovery of recombinant types could be conveniently reduced (or, of course, the same population size could provide more recombinants).

A common wheat cv. Chinese Spring "double mutant" is therefore being synthesized. At the moment the line actually carries a 5B chromosome with the ph1b mutation but, instead of bearing a 3D (ph2) chromosome, lacks the entire 3DS arm. Such a line was first developed instead of a euploid double mutant because the 3DL telocentric would have worked as a cytological marker, giving assurance of the actual absence of both Ph1 and Ph2 suppressors in the event of no detectable additive effect of the two (Table 3).

Pairing is being evaluated in the ph1b/3DS line as well as in the first hybrid combinations so far obtained between this genotype and more or less closely related species to wheat, such as Ae. variabilis and Secale cereale to start with. Various other hybrid combinations are planned. The ph1b/3DS line is being used in appropriate crosses to derive a true ph1+ph2 euploid mutant, also to possibly get an increase in self-fertility of such a mutant combination.

Table 1. Transfer of mildew resistance induced by use of the phlb mutation between Triticum longissimum (Ae. longissima) 3S^1 short arm chromosome and its group-3 wheat homoeologues: isolation of mildew resistant wheat recombinant types.

GROUP-3 WHEAT MONOSOME	N° PLANTS TESTED FOR MILDEW RES.	MILDEW RESISTANT PLANTS TOTAL N°	% TESTED OFFSPRING	RECOMBINANTS(a) TYPE N° and PROXIMAL	DISTAL	% of TOTAL PLANTS
3A	85	17	20.0	6 (2n=40,41,42)	1 (2n=39+t)	8.2
3B	58	9	15.5	2 (2n=40,41)	1 (2n=39+t)	5.2
3D	26	5	19.2	1 (2n=42)	1 (2n=39+t)	7.7

(a) Mildew resistant heterozygous recombinant plants isolated after cross with the euploid (cv. Chinese Spring, 2n=42) of $_1$plants where homoeologous pairing had been induced (2n=40+t= 19" + 1'3A(or 3B or 3D) + t'3S s + 1'5Bphlb).

Table 2. Male transmission of the transfer chromosome and meiotic pairing pattern exhibited by mildew resistant recombinant lines (T. longissimum 3S1s/group-3 wheat chromosomes) in BC generations to normal Chinese Spring.

RECOMB. LINE	1st GENERATION			2nd GENERATION		
	Offspring n°	Transm. %	M_I pairing (2n=42)	Offspring n°	Transm. %	M_I pairing (2n=42)
R1A	75	46.6	21"/19"+1'"+1'	-	-	-
R2A	21	38.1	19"+1'ᵛ/20"+2'	82	29.3	21"(a)
R3A	56	27.5	20"+2'/19"+1'"+1'	81	40.7	20"+2'/19"+1'"+1'
R4A	33	36.3	21"/20"+2'	46	28.3	21"(a)/20"+2'
R5A	32	65.6	21"/20"+2'/19"+1'ᵛ	92	31.5	21"(a)
R6A	27	40.7	19"+1'"+1'/20"+2'	83	31.3	21"(a)/19"+1'ᵛ
R1B	52	34.4	21"/20"+2'	58	32.8	21"/20"+2'
R2B	57	24.2	21"/19"+1'"+1'	121	27.3	21"/19"+1'"+1'
R1D	110	36.4	19"+1'"+1'/20"+2'	88	27.3	19"+1'"+1'/20"+2'

(a) All ring bivalents in some cells.

Table 3. Homoeologous pairing and recombination induced between different wheat and related alien chromosomes using chromosomal or genic mutations of the 5B wheat pairing control system.

Alien species	Pairing induction condition	Critical homoeologous chromosomes	Pairing frequency(%) (M_I records)(a)	Gametic recombination frequency(%)	References
Agropyron elongatum	nulli-5B/tetra-5D	3D/3Ag	10	6.0	2, 7, 8
	"	7D/7Ag	10	8.7	2, 7, 8
	"	3D/3AgL	–	2.1	9
A. elongatum	nulli-5B/tetra-5D	6D/6Ag	4.9	–	11
	mono-5Bph1b	"	2.5	–	11
A. intermedium	mono-5Bph1b	4B/Ai	2.8	–	10
	5Bph1b/ph1b	"	10.8	–	10
	"	4B/AiS	17.8	–	10
Ae. longissima	mono-5Bph1b	3A/3S^{1}S	8.4(b)	8.2	3, 15
	"	3B/3S^{1}S	11.6(b)	5.2	3, 15
	"	3D/3S^{1}S	8.3(b)	7.7	3, 15
Secale cereale	5Bph1b/ph1b	1D/1DS-1RL transl.	–	1.4	19

(a) These values, based on unavoidably rough M_I observations (see text), in most cases turned out to be underestimates of the actual pairing frequency (four times the gametic recombination frequency).

(b) These values refer to pairing of the alien telo (3S^{1}S) with its monosomic wheat homoeologue (indicated in the preceding column) and, though to a much lower extent, with either of the two disomic ones, |3|.

Figure 1 . Crossing scheme adopted to develop a common wheat cv. Chinese Spring (CS) mutant line lacking simultaneously the two main homoeologous pairing suppressor genes in wheat and wheat hybrids, Ph1 (chromosome 5BL) and Ph2 (chromosome 3DS).

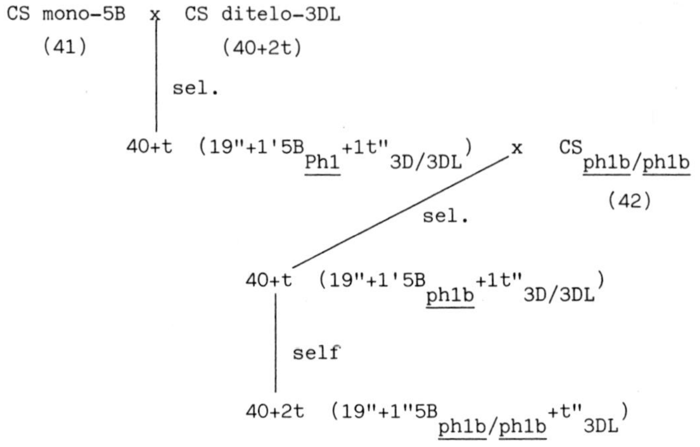

CS mono-5B   x   CS ditelo-3DL
(41)   (40+2t)

| sel.

40+t   (19"+1'5B$_{Ph1}$+1t"$_{3D/3DL}$)   x   CS$_{ph1b/ph1b}$

                                     (42)

sel.

40+t   (19"+1'5B$_{ph1b}$+1t"$_{3D/3DL}$)

| self

40+2t   (19"+1"5B$_{ph1b/ph1b}$+t"$_{3DL}$)

REFERENCES

|1| MUJEEB-KAZI, A. and D.C. JEWELL, CIMMYT's wide cross program for wheat and maize improvement, Proc. of the Inter-Center Seminar on Intern. Agric. Res. Centers (IARCs) and Biotechnology, 23-27 April 1984, Int. Rice Res. Inst., Manila, Philippines (1985) 219-226.

|2| SEARS, E.R., Chromosome engineering in wheat, Stadler Genet. Symp. 4 (1972) 23-38.

|3| CEOLONI, C., Transfer of alien genes into cultivated wheat and triticale genotypes by the use of homoeologous pairing mutants, Proc. 4th FAO/IAEA Research Coordination Meeting on: "Evaluation of semidwarf cereal mutants for cross breeding", Roma 16-20 December 1985 (1987) in press.

|4| SEARS, E.R., An induced mutant with homoeologous pairing in common wheat, Can. J. Genet. Cytol. 19 (1977) 585-593.

|5| GIORGI, B., A homoeologous pairing mutant isolated in Triticum durum cv. Cappelli, Mut. Breed. Newsl. 11 (1978) 4-5.

|6| GIORGI, B. and C. CEOLONI, A ph1 hexaploid triticale: production, cytogenetic behaviour and use for intergeneric gene transfer, Proc. Eucarpia Meeting on "Genetics and breeding of triticale", Clermont-Ferrand, France, July 1984 (1985) 105-117.

|7| SEARS, E.R., Agropyron-wheat transfers induced by homoeologous pairing, Proc. 4th Int. Wheat Genet. Symp., Columbia, Missouri(1973) 191-199.

|8| SEARS, E.R., Analysis of wheat-Agropyron recombinant chromosomes, Proc. 8th Eucarpia Congress on: "Interspecific hybridization in plant breeding", Madrid, Spain, 1977 (1978) 63-72.

|9| SEARS, E.R., Transfer of alien genetic material to wheat, In "Wheat science -today and tomorrow" (L.T. EVANS and W.J. PEACOCK Eds.), Cambridge Univ. Press (1982) 75-89.

|10| WANG, R.C., LIANG, G.H. and E.G. HEINE, Effectiveness of ph gene in inducing homoeologous chromosome pairing in Agrotricum, Theor. Appl. Genet. 51 (1977) 139–142.

|11| YASUMURO, Y., MORRIS, R., SHARMA, D.C. and J.W. SCHMIDT, Induced pairing between a wheat (Triticum aestivum) and an Agropyron elongatum chromosome, Can. J. Genet. Cytol. 23 (1981) 49–56.

|12| KIBIRIGE-SEBUNYA, I. and D.R. KNOTT, Transfer of stem rust resistance to wheat from an Agropyron chromosome having a gametocidal effect, Can. J. Genet. Cytol. 25 (1983) 215–221.

|13| GIORGI, B. and F. BARBERA, Intergeneric hybrids, meiotic behaviour and amphiploids involving a ph mutant of Triticum turgidum ssp.durum, Proc. Symp. on "Induced Variability in Plant Breeding", Eucarpia, Wageningen (1981) 96–100.

|14| CEOLONI, C., GIORGI, B. and L. ROSSI, Research on triticale in Italy: results from conventional and novel breeding techniques, Proc. Int. Triticale Symp., Sydney, Australia (1986) 428–433.

|15| CEOLONI, C., Incorporation of a mildew resistance gene derived from T. longissimum into common wheat through crossing-over between homoeologous chromosomes, Proc. XXVIII Annual Meeting of the Italian Soc. for Agric. Genet., Bracciano, Rome (1984) 94–95.

|16| CEOLONI, C., Triticum longissimum chromosome G ditelosomic addition lines: production, characterization and utilization, Proc. 6th Int. Wheat Genet. Symp., Kyoto, Japan (1983) 1025–1031.

|17| CEOLONI, C. and G. GALILI, Chromosome arm location and mode of expression of a phosphodiesterase gene from diploid wheat Triticum longissimum, Cer. Res. Comm. 10 3–4 (1982) 151–157.

|18| DHALIWAL, H.S., GILL, B.S. and J.G. WAINES, Analysis of induced homoeologous pairing in a ph mutant wheat x rye hybrid, J. of Hered. 68 (1977) 207–209.

|19| KOEBNER, R.M.D. and K.W. SHEPHERD, Induction of recombination between rye chromosome 1RL and wheat chromosomes, Theor. Appl. Genet. 71 (1985) 208–215.

|20| SEARS, E.R., A wheat mutation conditioning an intermediate level of homoeologous chromosome pairing, Can. J. Genet. Cytol. 24 (1982) 715–719.

|21| SEARS, E.R., Mutations in wheat that raise the level of meiotic chromosome pairing, Proc. 16th Stadler Genet. Symp.(J.P. Gustafson Ed.), Plenum Press, New York and London (1984) 295–300.

MALE STERILE FACILITATED RECURRENT SELECTION

DUANE E. FALK

Crop Science Department
University of Guelph
Guelph, Ontario, Canada

M. Maluszynski (ed.), Current Options for Cereal Improvement, 31–38.
© *1989 by Kluwer Academic Publishers.*

32

Abstract

The use of recurrent selection in self-pollinated crops has
been limited due to the difficulty of making the large numbers
of crosses required for each recurrent cycle. The use of
genetic male sterility has helped to overcome this limitation
but there are some problems associated with the manipulation of
the male sterile character in populations. The linking of a
male sterile gene (msg6) with a shrunken endosperm gene (sex1)
which expresses xenia has created an efficient method of
manipulating the male sterile gene in barley. Seeds which will
produce male sterile plants can be identified before sowing
based on the shrunken endosperm and so can be planted in rows
or strips adjacent to any desired male parents for outcrossing.
Seeds which will produce only fertile plants can also be
identified and grown separately. Since the seeds which will
produce male sterile plants can be identified, they can be used
as females in controlled crosses where they produce more seeds
with much less effort than conventional emasculating and
crossing techniques. The shrunken endosperm seeds can also be
used to create composite crosses on a field scale. The use of
male sterility in recurrent selection for population breeding
and improvement can be combined with haploidy for rapid
fixation of lines with desirable alleles concentrated in them.
Male sterile facilitated recurrent selection can be readily
used in less developed countries to improve adapted landrace
germplasm. It can also be used to advantage in modern
industrialized countries where budget cuts have reduced the
amount of technical help available for such operations as
crossing. Since more crosses can be made using male sterile
plants, a greater range of variability should be available for
breeders to select from. The frequent recombining within
populations undergoing recurrent selection should lead to a
rapid accumulation of desirable alleles resulting in
potentially better cultivars.

Recurrent Selection

Recurrent selection has been used successfully in cross pollinated crops such as maize and many forage species. Dudley [1] explained the success of selecting for oil and protein in a population of Burr Leaming maize by noting that "mild selection combined with frequent opportunities for recombination have been powerful tools in achieving progress with a minimal number of observations." The real power of recurrent selection can be illustrated by examples of selection for chemical composition in maize. The highest protein content known in maize is in Illinois High Protein (23%), the highest oil content is in Illinois High Oil (15.5%), the lowest protein is in Illinois Low Protein (5%) and the lowest oil is in Illinois Low Oil (0.7%) - all four of these lines representing the extremes for the entire _Zea mays_ species originated from the same small population and were developed by recurrent selection [1].

Recurrent selection where selected individuals are intercrossed to form the base population for the next cycle of selection results in a higher proportion of the individuals in the succeeding population containing desirable alleles. More of these desirable alleles are also present in some individuals which should be selected in the next cycle. By continuing to select the individuals with the most desirable alleles and recombining them, it is theoretically possible to produce individuals with a high proportion of the total desirable alleles after several cycles without substantially reducing the variability for traits not under selection.

Male sterile facilitated recurrent selection

The use of recurrent selection in self-pollinated crops has been limited due to the difficulty in making the large numbers of crosses required for each recurrent selection cycle. Recurrent selection has been applied to self-pollinated crops by modifying the procedure itself [2] or by using special techniques, such as haploidy [3]. Genetic male sterility has been employed in conjunction with recurrent selection in soybeans [4] and in barley [5].

The use of genetic male sterility in self pollinated crop helps to overcome the limitation on the numbers of crosses that can be produced by eliminating the necessity of emasculation. However, identifying the male sterile plants prior to flowering is a problem, especially where selection for other traits, with subsequent intercrossing, is being practiced in the population. The male sterile plants are often pollinated by undesired pollen before their identity as male sterile plants can be established, thus reducing the effectiveness of the recurrent selection system.

Several schemes have been proposed to identify male sterile plants prior to flowering [6,7,8,9]. The system based on linkage of a xenia-expressing shrunken endosperm gene with a male sterile gene [10] has been used in a number of programs and is simple and reliable.

The close linkage of the msg6 locus to the sex1 locus on barley chromosome 6 has made it possible to manipulate the male sterile gene as if it also expressed xenia. The coupling linkage of these two genes allows the identification of seeds with a very high probability of being homozygous for the msg6 alleles prior to planting simply by selecting shrunken seed (sex1 sex1) in the progeny of a heterozygous plant (Figure 1). The shrunken seeds (which will produce male sterile plants) can be sown as rows or plots adjacent to any other barley to achieve cross pollination on a field scale.

Selection of plump seeds (Sex1 --) will give populations of normal, male fertile plants which can be used as pollen sources or to produce selfed seed. Fertile plants with all plump seed (Sex1 Sex1) will be homozygous for the male fertile allele (Msg6 Msg6) and so can be handled like any normal barley breeding line. Fertile plants segregating for shrunken seed (Sex1 sex1) are heterozygous; the shrunken seed will give male sterile plants which can be used for further crossing, the plump seed will give fertile plants which will self-pollinate and can be used as pollen doners. Since the linkage between sex1 and msg6 is very tight (< 1% recombination), the presowing selection system is conserved in the population in all the heterozygous plants [10].

Since the seeds which will produce male sterile plants can be readily identified prior to sowing, they can be used for controlled crosses as well as field-scale crossing. Male sterile plants generally set more seed from crossing than emasculated plants because there is no damage to the stigma during anther removal. It is also possible to cross onto numerous tillers of a single male sterile plant as they tend to tiller more profusely than self-fertile plants. Crosses have been made onto as many as 25 tillers of a single male sterile two-rowed plant with an average of 22 seeds per crossed spike.

When heterozygous plants are grown in a good environment, subsequent separation of plump and shrunken seeds is quite easy. A good technician familiar with the shrunken endosperm trait can separate shrunken seeds from plump at a rate of about 1500 per hour. Thus a large number of potentially male sterile plants can be identified in a short time. Since large numbers of female plants can be grown, it is possible to create large amounts of crossed seed in most environments. In some environments, such as California and Arizona winter nurseries with irrigation, very high seed sets can be obtained. In 1979-80, 800 male sterile plants produced over 20,000 outcrossed seed in a California winter nursery by natural crossing.

Numerous male parents can be used to synthesize a population containing several genes affecting a single trait, such as disease resistance [5], which can then be used as a base population for recurrent selection. If selection can be practiced prior to flowering, then alternating rows of male sterile and fertile plants can be grown for maximum recombining subsequent to selection. If selection cannot be made until after flowering, then only plump seed would be planted and all selected plants would be fertile and produce selfed progeny. Shrunken seed from selected plants could be used as the females for the next cycle of intercrossing with plump seed from other selected plants being used as the pollen source.

It would be possible to use recurrent selection techniques for yield by planting only plump seed (which would produce all fertile plants) in a normal yield plot then select shrunken or plump seed from the high yielding plots for intercrossing in the following generation. If all plots were of the same generation and originated from heterozygous plants, they would all produce the same proportion of shrunken seed and so no correction for shrunken seed would need to be applied.

Recurrent Selection and Haploidy

The use of haploid procedures such as the Bulbosum method [11] to fix the alleles in lines selected from a recurrent selection population would make it possible to monitor progress due to selection as well as provide materials that may be tested for cultivars on their own merit. By selecting plants producing only plump seed (Sex1 Sex1) for use as a source of haploids, all of the resulting doubled haploids should be fully fertile and have normal seed. These lines could be increased and tested for a number of traits. Since male sterile plants cannot be maintained or increased through self-pollination, there is no advantage to using lines that could produce some male sterile progeny in the haploid generation.

Recurrent Selection in the Third World

This simple, conserved method of obtaining high proportions of male sterile plants for breeding purposes makes it feasible to use recurrent selection as a method of introgressing desirable agronomic traits into adapted germplasm on a large scale. If a shrunken endosperm male sterile stock is used as the female in repeated backcrosses to lines carrying desirable agronomic traits, it is possible to develop a near-isogeneic population. These can then be used in crosses on a field scale with broadly-based, adapted germplasm. Such a scheme could be used to incorporate desirable agronomic characters such as lodging resistance and early maturity into landraces in the Third world without changing the basic adaptation nor narrowing the genetic base of the landrace population to any appreciable degree. The variability of the "recurrent" parent germplasm may even be increased because each intercross cycle different

36

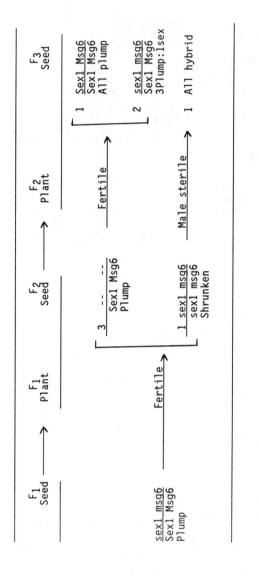

FIGURE 1. Segregation and phenotypic expression of sex1 msg6 in advancing generations (-- = either genotype).

genotypes from the heterogeneous landrace may be used as pollen doners. If the crossing is done on a field scale with good seed set on the male sterile plants, it should provide highly variable populations for selection while maintaining most of the desirable adaptation of the original landraces.

Since a minimum of technical training is required to identify the shrunken seeds, using the male sterile plants grown from them as female parents in crosses should be well within the capability of even the most modest breeding programs in the world. By obtaining a few shrunken seeds and making some crosses of local material onto them, it should be possible to develop well-adapted populations of good breeding material in a few years. With some selection and further crossing (onto male sterile plants), a sound breeding program, using primarily local materials, should result and superior, new cultivars should be released. If the populations are grown in the local environments and seed from male sterile plants harvested it should be possible to accumulate a number of desirable alleles for adaptation and tolerance to the normal stresses of the region. By promoting an exchange of genetic material among local populations, new recombinations may be found which are superior to any of the existing genotypes.

Summary

The use of recurrent selection is possible in self-pollinated crops when mechanisms such as genetic male sterility are employed. Genetic male sterility can be used more efficiently when combined with a xenia-expressing trait such as the shrunken endosperm in barley. A system has been developed based on the tight linkage of msg6 and sex1 on barley chromosome 6. This system can be used for presowing selection of male sterile plants in recurrent selection populations. The system is simple to use, it is conserved within the populations developed from it and it is simple to eliminate the male sterility from the population when desired. A germplasm containing this system is publicly available.

Literature Cited

[1] Dudley, J.W. R.J. Lambert, and D.E. Alexander, "Seventy generations of selection for oil and protein concentration in the maize kernel", In: Seventy Generations of Selection for Oil and Protein in Maize. J.W. Dudley, Ed. Crop Science Society of America, Madison, Wisconsin, U.S.A. (1974).

[2] Compton, W.A., Recurrent selection in self-pollinated crops without extensive crossing. Crop Sci. 8 (1968) 773. 1968.

38

[3] Patel, J.D., E. Reinbergs and S.O. Fejer, Recurrent
     selection in doubled-haploid populations of barley
     (Hordeum vulgare L.). Can. J. Genet. Cytol. 27 (1985)
     172-177.

[4] Brim, C.A. and C.W. Stuber, Application of genetic male
     sterility to recurrent selection schemes in soybeans.
     Crop Science 13 (1973) 528-530.

[5] Sharp, E.L., "Changing gene frequencies in populations,"
     In: Durable Resistance in Crops. Ed. F. Lamberti,
     J.M. Wallen and N.A. Van der Graaff. Plenum
     Publishing Corp. (1983).

[6] Wiebe, G.A., A proposal for hybrid barley. Agron. J. 52
     (1960) 181-182.

[7] Eslick, R.F., "Balanced male sterile and dominant
     preflowering selective genes for use in hybrid seed
     production", In: Barley Genetics II, Proc. 2nd Int.
     Barley Genet. Symp., Pullman, WA. (1970).

[8] Foster, C.A., A balanced male sterile chloroplast mutant
     scheme for hybrid barley. Barley Genet. Newsl. 9
     (1979) 22-23.

[9] Falk, D.E., K.J. Kasha and E. Reinbergs, "Presowing
     selection of genetic male sterile plants to facilitate
     hybridization in barley," In: Barley Genetics IV,
     Proc. 4th Int. Barley Genet. Symp., Edinburgh, U.K.
     (1981).

[10] Falk, D.E., M.J. Swartz and K.J. Kasha, Linkage data with
      genes near the centromere of barley chromosome 6.
      Barley Genet. Newsl. 10 (1980) 13-16.

[11] Kasha, K.J. and K.N. Kao, High frequency of haploid
      production in barley (Hordeum vulgare). Nature
      (Land). 225 (1970) 874-876.

WIDE HYBRIDIZATION FOR CEREAL IMPROVEMENT

G. FEDAK

Plant Research Centre
Research Branch, Agriculture Canada
Ottawa, Ontario K1A OC6
Canada

M. Maluszynski (ed.), Current Options for Cereal Improvement, 39–48.
© *1989 by Kluwer Academic Publishers.*

ABSTRACT

An overview is presented of the extent of wide hybridization within the Triticeae concentrating on wheat, barley and rye. The potential for gene transfer is indicated, some examples are given of practicality of realized hybrids and the scope for further studies on existing hybrids and possibilities for additional combinations.

INTRODUCTION

Heritable genetic variation is the essence of plant improvement. The task of the plant breeder-geneticist is to accumulate the variation and then to manipulate it for the improvement of a particular crop plant. When the variability inherent within a species has been exhausted or is nonexistent, the breeder must examine related species and genera. Because of the problems inherent in intergeneric hybridization such as low crossability, sterility of hybrids, resistance to chromosome doubling, linkage to undesirable traits, extremely long-term and labor intensive manipulations not to mention that of dilution of elite germplasm, the choice of parents from wild species and genera is done as a last resort.

Ongoing breeding programs in most crop plant species are making steady, incremental improvements in most important traits. However, even after extensive screening of cultivars and land races, some traits are lacking in creal crops in particular. For example, it was found after extensive screening (A. Comeau, personal communication) that neither Triticum nor Aegilops species have resistance to Barley yellow Dwarf Virus (B.Y.D.V.). Fusarium graminearum resistance was thought to be in the same category until resistance was detected recently in introductions of wheat from the People's Republic of China (J.D. Miller, p.c.). Similarly, barley does not have adequate resistance to Helminthosporium sativum. In addition, both crops could be improved in terms of tolerance to drought, salinity and low temperatures and there is negligible variability for these traits within the crop. Any additional variability is a bonus in any breeding program permitting more flexibility and options and most breeders, even the very best will admit that they probably aren't utilizing the variability that is available within a species let alone related genera.

Barley-wheat hybrids

The interest in a hybrid between wheat and barley is the combination of traits between two crop species and thus avoid the dilution of elite germplasm caused by crossing a crop plant with a wild-weedy relative. The first reported attempt [10] ended in failure but by employing more elegant techniques, Kruse [22] was able to produce the first hybrids, and these results stimulated additional efforts. Since that time, numerous hybrids have been produced between cultivars or species of the two genera. In a recent review by Fedak [11], it was shown that a total of 32 hybrids have been produced from barley cultivars as maternal parents and hexaploid wheat cultivars but only six from the reciprocal cross. Crossing of diploid and tetraploid wheats onto barley cultivars has yielded 13 hybrid combinations. The intercrossing of species of one genus with species/cultivars of the other has produced six hybrid combinations; four with Triticum as the maternal parent. Since that review was written additional hybrid combinations have been produced in each category.

One of the potential practical applications of these hybrids is the transfer of BYDV resistance from barley to wheat. The resistance gene Ryd$_2$ is located on chromosome 3 in barley [28]. The resistance gene is not carried by all cultivars but is present in the strain CI2376. We have crossed this strain onto wheat cultivar Chinese Spring, obtained backcross progenies and eventually some monosomic addition lines, one of which represented barley chromosome 3. The infection of that line with viruliferous aphids gave inconclusive results on resistance to BYVD. One reason for the lack of distinct results may have been the poor agronomic performance of the wheat cultivar Chinese Spring. Secondly, since all reported Triticum-Hordeum hybrids resemble the Triticum parent in morphology, this may be an indication of the suppression by the latter of other traits carried by the Hordeum parent. Ongoing experiments will indicate if the latter concept is valid.

Gene introgression from barley to wheat has already been reported with positive beneficial results. Complete hybrids were obtained between Chinese Spring wheat and 4x H. bulbosum. The sterile F$_1$ was backcrossed to Chinese Spring monosomic 5B resulting in six addition lines, two of which had protein contents of 22.30 and 20.37%. In addition, the progeny of two backcross-derived euploid plants were completely resistant to wheat yellow mosaic virus (WYMV) [30].

Fertile amphiploids have been produced from hybrids between H. chilense and several strains of T. turgidum [23, 8]. Although the grain yield of these tritordeums is lower than that of the wheat parent, the protein content is unusually high, up to 25.2% in some combinations compared to the wheat parent at 13.8%. We have obtained the amphiploid from the H. californicum x Chinese Spring hybrid [16], thus an octoploid tritordeum. This version is also a vigorous strain with near perfect fertility. It has not been characterized for agronomic traits but reciprocal backcrosses have been made to Chinese Spring (plus to rye and other amphiploids) and the production of addition lines of H. californicum has been initiated.

Both of the tritordeum strains noted above are nearly fully fertile, with well-filled though somewhat smaller kernels than the wheat parent, i.e. much more impressive than any primary triticale. A breeding program could be initiated in either one by first broadening the genetic base. A careful and systematic selection for desirable traits in parental strains would provide a solid basis for a breeding program in a new crop.

Wheat-Agropyron

Numerous hybrids have been made between wheat and Agropyron species over the years [25, 27, 5, 17, 7]. The Agropyron species are known to carry traits such as resistance to BYDV and wheat streak mosaic virus (WSMU) [28]; rust and mildew resistance [21] and salt tolerance [9]. Species of the genus Leymus are also extremely hardy perennials. More specifically, Leymus angustus, which has been crossed onto cultivars Fukuhokomugi and Asakazekomugi of bread wheat [7] and onto T. turgidum [Fedak, unpublished] has better tolerance to ice encasement than Secale cereale or A. elongatum [20] and also has resistance to snow mould (Tifula sp.) and BYDV (Comeau, p.c.). Backcrossing and attempts at tissue culture regeneration are underway in an effort to transfer as many useful traits as possible.

Agropyron repens is an extremely persistent perennial and for that reason one of the more troublesome weeds in Canadian agriculture. Numerous attempts have been made in the past to cross poorly defined "A. repens" accessions onto wheat with negligible success. More recently in hybrids with bread wheat no homology was detected between parental genomes [15], and self-sterility was overcome by crossing back to wheat (Comeau, p.c.). No systematic screening or evaluation has been performed on the backcross progenies but a wide range of variability exists for such traits as plant height, straw strength and kernel size.

Triticale in wide crosses

The genetic variability present in Agropyron species can also be used for triticale improvement. The first step in this direction was achieved by the crossing of several strains of Agropyron onto triticale. Three strains of A. intermedium were crossed onto Welsh triticale [18] resulting in the production of hyrids at frequencies of 1.6 to 2.6% of pollinted florets. More recently, A. junceum 6x and A. elongatum 2x were crossed onto triticale cultivars Armadillo and Carman (Fedak, unpublished).

Triticale has been crossed quite readily onto barley cultivars. Three trigeneric hybrids were produced by crossing 6x and 8x triticale onto Elgina barley [6] and another six between a number of spring cultivars of each species [2]. Triticale cultivars Rosner, DRIRA and Welsh have been crossed onto H. parodii 6x [19] and Beagle onto H. pubiflorum (Fedak, unpublished). Backcrossing of these hybrids has not been initiated as yet.

Wheat-rye crosses

The genus Secale has already made noteable contributions to wheat improvement as one component in triticale and secondly by contributing disease resistance carried on translocated chromosomal segments or chromosome substitutions. Rye has already contributed leaf, stem, stripe rust and mildew resistance to numerous wheat cultivars through chromosomal 1B-1R substitutions or 1B-1R translocations [24, 32].

In 1980, a rather severe attack of Fusarium graminearum (scab) occurred in Ontario winter wheat. At that time, no sources of resistance had been identified in wheats but resistance was reported in land races of rye in Brazil [1]. These were crossed onto a number of bread wheat cultivars and the octoploid triticales were produced (unpublished). A total of 30 strains were inoculated with the toxin isolated from the fungus by Dr. J.D. Miller and six lines were found to be resistant to the toxin. These will be backcrossed to wheat in an effort to transfer the resistance.

Barley in wide crosses

In crosses with Triticum species, Hordeum species have generally been the donors in terms of chromosome addition lines and introgression of agronomic traits. Wide crosses of barley to other Hordeum species have not been as successful as wheat crosses to Triticum or Aegilops species. Although there are 26 Hordeum species in total [3] their crossability with barley is very low [4] and chromosome pairing in interspecific hybrids with the exception of H. bulbosum and H. spontaneum, is negligible. Nevertheless, many of the Hordeum species carry useful agronomic traits and some of these such as disease resistance and stress tolerances have been expressed in hybrids or backcross progenies (see review by Fedak [11].

In terms of intergeneric hybrids, many more combinations with Secale, Agropyron and Elymus have been made with Hordeum species than with H. vulgare. For example, a total of 28 hybrids have been obtained between Secale spp. and/or cultivars and Hordeum species but only 21 onto H. vulgare; 20 combinations of Elymus species onto Hordeum species but only 9 combinations with H. vulgare and 13 Agropyron species with Hordeum species but only 5 with H. vulgare. (Most of the above hybrid combinations are listed in the review by Fedak [11]. In the latter hybrids, the species A. intermedium [12], A. caninum and A. dasystachyum [13], A. junceum 6x and A. elongatum 2x (Fedak, unpublished) were successfully utilized to produce hybrids. A hybrid between Betzes barley and Hystrix patula 4x was pistilloid and therefore not studied any further.

A prevalent fungal disease of barley in Ontario Canada for which there is no adequate resistance is the root rot-leaf spotting complex caused by the fungus Helminthosporium sativum. Some of the intergeneric hybrids described above were artificially inoculated with the fungus by Dr. R.V. Clark, Plant Pathologist at the Plant Research Centre, Agriculture Canada, Ottawa. The leaves of the hybrid between Betzes and A. caninum were virtually free of lesions so this hybrid deserves more consideration in further studies to improve the disease resistance of cultivated barley.

Rye (Secale cereale L.) is probably the most hardy and persistent of the cereal crops with tolerance/resistance to most diseases and tolerance to most climatic and biotic extremes. Its contribution to wheat improvement have been substantial. The same attributes could have a marked influence on barley improvement. One of our goals was to attempt to improve the winter hardiness of barley to make it suitable for Canadian climatic conditions. As pointed out above, numerous crosses have been attempted with these objectives in mind. A total of 16 hybrid combinations were reported by eight different authors [11] employing H. vulgare as the maternal parent. The type of progeny obtained included necrotic seedlings, subviable vegetative clumps, barley haploids and complete or incomplete hybrids. The amphiploids that were produced were sterile because of chromosome instability.

Hordeum-Secale hybrids are easier to produce by employing wild species of Hordeum and more hybrids have resulted from that approach including one partially fertile amphiploid. This one originated from a hybrid between H. pubiflorum and S. africanum [14]. It was perennial in growth habit with a brittle rachis and predominantly Secale morphological features on the spike. It was partially fertile producing badly shrivelled seeds some of which were viable. This amphiploid may now be employed as a bridge to transfer Secale traits to barley.

Other hybrid combinations reported to have some degree of self fertility were those between H. jubatum 4x and S. cereale 4x [31] and secondly between H. geniculatum 4x and S. cereale 4x [26].

CONCLUSIONS

In the above discussion, many of the hybrid combinations mentioned are, with the exception of the tritordeums, at a sterile F_1 hybrid stage. Extensive backcrossing will be required to transfer the desirable traits into a stable improved plant type that will be directly useable by the plant breeder. The backcrossing phase is one of the weakest links in the process of transferring genetic traits from an F_1 intergeneric hybrid to useable germplasm. Some hybrids are female sterile and do not respond to backcrossing. For those that respond the backcrossing procedure is very time consuming and labor intensive and often involves selection for fertile and chromosomally stable individuals. In most cases, neither the cytogeneticist nor breeder have the extra "pairs of hands" to handle this aspect of wide crossing.

Nevertheless, considering the number of intergeneric hybrids that have already been made among species within the tribe Triticeae and the large numbers of species that have yet to be exploited within the genera of Agropyron, Elymus, Psathyrostachys, etc., the potential for crop improvement through wide crosses is quite extensive.

REFERENCES

[1] BAIER, A.C., DIAS, J.C.A., NEDEL, J.L., (1980)
    Triticale Research. Annual Wheat Newsletter,
    26: 47.

[2] BALYAN, H.S., FEDAK,G., Meiotic study of hybrids
    between barley (Hordeum vulgare L.) and triticale
    (XTriticosecale Wittmack). (in preparation).

[3] BOTHMER, R. von, JACOBSEN, N., JORGENSEN, R., (1981)
    Phylogeny and taxonomy in the Genus Hordeum.
    Barley Genetics IV. Proc. 4th Int. Barley
    Genetics Symp. Edinburgh, 13.

[4] BOTHMER, R. von, FLINK, J., JACOBSEN, N.,
    KOTIMAKI, M., LANDSTROM, T., (1983) Interspecific
    hybridization with cultivated barley (Hordeum
    vulgare L.), Hereditas, 99: 219.

[5] CAUDERON, Y., (1979) Use of Agropyron species for
    wheat improvement. Proc. Conf. Broadening
    Genetic Base of Crops, Wageningen, 175

[6] CLAUSS, E. (1980) Trigeneric hybrids between barley,
    wheat and rye. Cereal Res. Comm. 8: 341.

[7] COMEAU, A., FEDAK, G., ST. PIERRE, C.A.,
    THERIAULT, C., (1985) Intergeneric hybrids between
    Triticum aestivum and species of Agropyron and
    Elymus. Cereal Res. Comm. 13: 149.

[8] CUBERO, J.I., MARTIN, A., MILLAN, T., GOMEZ-
    CARERA, A., DE HARO, A., (1986) Tritordeum: A
    new alloploid of potential importance as a
    protein source crop. Crop Sci. 26: 1186.

[9] DVORAK, J., ROSS, K., MEDLINGER, S., (1985) Transfer
    of salt tolerance from Elytrigia pontica (Podp.)
    Holub to wheat by the addition of an incomplete
    Elytrigia genome, Crop Sci. 25: 306.

[10] FARRER, W., (1904) Some notes on the wheat "Bobs",
    its peculiarities, economic value and origin.
    Agric. Gazette of N.S.W., 15: 849.

[11] FEDAK, G., (1985) Wide crosses in Hordeum. Barley,
    Agronomy Monograph Number 26. American Society of
    Agronomy, Crop Science Society of America, Soil
    Science Society of America, 156.

[12] FEDAK, G., (1985) Intergeneric hybrids between
    Hordeum vulgare and Agropyron intermedium var.
    trichophorum. Z. Pflanzenzüchtg. 95: 45.

[13] FEDAK, G., (1985) Intergeneric hybrids of Hordeum vulgare with Agropyron caninum and A. dasystachyum. Can. J. Genet. Cytol. 27: 387.

[14] FEDAK, G., (1985) Cytogenetics of a hybrid and amphiploid between Hordeum pubiflorum and Secale africanum. Can. J. Genet. Cytol. 27: 1.

[15] FEDAK, G., COMEAU, A., ST. PIERRE, C.A. (1986) Meiosis in Triticum aestivum x Elytrigia repens hybrids. Can. J. Genet. Cytol. 27: 430.

[16] GUPTA, P.K., FEDAK, G., (1985) Intergeneric hybrids between Hordeum californicum and Triticum aestivum. J. Hered. 76: 365.

[17] GUPTA, P.K., FEDAK, G., (1986) Hybrids of bread wheat (Triticum aestivum) with Thinopyrum scirpeum (4x) and Thinopyrum junceum (6x). Plant Breeding, 97: 107.

[18] GUPTA, P.K., FEDAK, G., (1986) Intergeneric hybrids between XTriticosecale cv. Welsh (2n=42) and three genotypes of Agropyron intermedium (2n=42). Can. J. Genet. Cytol. 28: 176.

[19] GUPTA, P.K., FEDAK, G., (1986) Variation in the induction of homoeologous pairing among chromosomes of 6x Hordeum parodii as a result of three triticale (XTriticosecale Wittmack) cultivars. Can. J. Genet. Cytol. 28: 420.

[20] HOPE, H.J., COMEAU, A., HASTY, P., (1984) Ice encasement tolerance of Prairieland wild ryegrass, Orbit tall wheatgrass and Puma rye grown under controlled environments. Cereal Res. Com. 12: 101.

[21] KNOTT, D.R., DVORAK, J., (1976) Alien germplasm as a source of resistance to disease. Ann. Rev. Phytopathology, 14: 211.

[22] KRUSE, A., (1973) Hordeum x Triticum hybrids. Hereditas, 73: 157.

[23] MARTIN, A., LAGUNA, E.S., (1982) Cytology and morphology of the amphiploid Hordeum chilense x Triticum turgidum conv.durum. Euphytica, 31: 261.

[24] METTIN, D., BLUTHNER, W.D., SCHLEGEL, G. (1973) Additional evidence on spontaneous IB/IR wheat-rye substitutions and translocations. Proc. 4th Int. Wheat Genet. Symp., Columbia, 179.

[25] MUJEEB-KAZI, A., ROLDAN, S., MIRANDA, J.L., (1984)
     Intergeneric hybrids of Triticum aestivum L. with
     Agropyron and Elymus species. Cereal Res. Comm.
     12: 75.

[26] PERSHINA, L.A. SHUMNY, V.K., BELOVA, L.I.,
     NUMEROVA, O.M., (1985) Hordeum geniculatus All x
          Secale cereale L. hybrids and their backcross
     generations with rye. Cereal Res. Comm. 13: 141.

[27] SHARMA, H.C., GILL, B.S. (1982) Current status of
     wide hybridization in wheat. Euphytica, 32: 17.

[28] SHARMA, H.C., GILL, B.S., UYEMOTO, J.K. (1984) High
     levels of resistance in Agropyron species to
     barley yellow dwarf and wheat streak mosaic
     virus. Phytopath. Z. 110: 143.

[29] TSUCHIYA, T., (1985) Linkage maps of barley. Barley
     Genetics Newsletter, 15: 82.

[30] WANG, L., ZHU, H., GUAN, Q., RONG, J., (1986)
     Production of T. aestivum (6x) - H. bulbosum (4x)
     alien disomic addition lines and introgression of
     resistant gene (WYMV) from H. bulbosum to common
     wheat. Proc. Fifth Int. Barley Genet. Symp.,
     Okayama, 94.

[31] WOJCIECHOWSKA, B., (1984) Crosses of barley with
     rye, Hordeum jubatum x 4x Secale cereale and BC
     progenies of H. jubatum x 2x S. cereale. Cereal
     Res. Comm. 12: 67.

[32] ZELLER, F.J., (1973) IB/IR wheat-rye chromosome
     substitutions and translocations. Proc. 4th Int.
     Wheat Genet. Symp., Columbia, 209.

THE EFFECT OF DWARFING GENES ON THE EXPRESSION OF HETEROSIS FOR GRAIN YIELD IN F_1 HYBRID WHEAT

M.D. GALE, A.M. SALTER AND W.J. ANGUS*

Plant Breeding Institute
Maris Lane, Trumpington
Cambridge CB2 2LQ, UK

*Nickerson RPB Ltd.,
Rothwell
Lincoln LN7 6DT, UK

M. Maluszynski (ed.), Current Options for Cereal Improvement, 49–61.
© *1989 by Kluwer Academic Publishers.*

ABSTRACT

Thirty six F_1 hybrids were produced with the use of the Shell chemical hybridising agent, WL 84811. Three tall male parents were each hybridised with four lines in the same three varietal backgrounds which were isogenic for the rht (tall), Rht1 and Rht2 (semi-dwarf) and Rht3 (dwarf) alleles at the GA-insensitive dwarfing gene loci. A yield trial of the F_1s and the parents allowed comparisons to be made between the effects of each dwarfing allele as a heterozygote, in both isogenic and intervarietal hybrids. The highest levels of heterosis were found in Rht3/rht genotypes. Although the actual yields of the Rht2/rht F_1 hybrids were marginally higher, the additional straw strength of Rht3/rht F_1s may, in high yielding environments, make them the Rht genotype of choice for F_1 hybrid breeders. Other results from this experiment included the observation that the use of WL 84811 itself has no effects on yield.

1. INTRODUCTION

At a previous FAO/IAEA Research Coordination meeting we described our first experiments to evaluate the use of Rht3, the Tom Thumb dwarfing gene, in chemical hybridising agent (CHA) produced F_1 hybrid wheat [1]. Briefly stated, the results indicated that (i) Rht3/rht heterozygotes, in an isogenic background, are of ideal semi-dwarf phenotype and have a grain yield potential similar to comparable talls, (ii) these yields derive, however, from more, but smaller, grains on a similar number of ears, (iii) heterosis for yield, where present in intervarietal F_1s, is expressed as increased grain weight with minimal effects on grain numbers or tiller density, and (iv) the use of Rht3 in intervarietal F_1s effectively channels heterosis into grain rather than vegetative growth, by providing more (+36%) grain per ear.

This report describes an experiment with a more comprehensive set of isogenic and intervarietal hybrids, designed to address the questions, (i) what are the relative effects of Rht1, Rht2 and Rht3 homozygotes and heterozygotes on yield in isogenic backgrounds?, (ii) are the levels of heterosis obtained in intervarietal F_1s affected by Rht genotype?, and (iii) if so, which genotype has the best potential for commercial F_1 hybrid wheat breeding? In addition we were able to test whether the CHA itself had any effects on the F_1 crop characteristics.

2. MATERIALS AND METHODS

2.1. Genotypes

The twelve parental genotypes used in the F_1 production blocks consisted of three sets of four near-isogenic lines, each carrying one of four alleles at the Rht loci on chromosomes 4A and 4D. The sets were in the varietal backgrounds of the European winter wheats, Maris Huntsman (H), Widgeon (W) and Bersee (B), The isogenic genotypes were rht (tall alleles at both loci), Rht1 (a 'Norin 10' allele on 4A), Rht2 (the other Norin 10 allele on 4D) and Rht3 (the 'Tom Thumb' dwarf allele at the 4A locus). The lines were produced by recurrent backcrossing to the tall variety, while holding the Rht allele heterozygous. After six backcrosses, rht and Rht homozygotes were obtained for each line by selfing. The rht lines used here are reconstituted talls obtained in this way.

2.2. F_1 grain production

Drilled plots, 6 m x 1.16 m, of the 12 female parents were sown inside three male (H, W and B) rht production blocks. The females were sprayed with the Shell CHA WL 84811, azetidine-3-carboxylic acid, at the optimum growth stage and dose rate.

F_1 grain yields varied between 1 kg and 4 kg per plot. Widgeon was the best male parent, with a mean of 2.5 kg per female plot, and Huntsman the worst, giving a mean of 1.3 kg. No differences in efficiency as females were seen between varietal backgrounds or Rht genotypes.

2.3. F_1 yield trial

The 36 F_1s and 12 inbred parents were trialled as 6 m x 1.16 m plots, randomised in four blocks. The trial was given the complete current NIAB/ADAS prophylactic fungicide. Herbicide and insecticide treatment was as recommended for on-farm use. Anthesis time (50% anther extrusion within the plot), ear density (estimated from a 30 cm strip across each plot) and height were measured before harvest. Twenty random shoots were cut at ground level from each plot before combine harvesting. These were analysed to assess grain number per ear, ear yield and harvest index. Grain weights were measured in a 200 grain sample taken from the whole plot grain. Yields are presented as $g.m.^{-2}$ over the measured area of the plot and no deflationary allowance has been made for edge effects. Lodging was not prevented and did occur in some of the taller plots,however this took place shortly before harvest and was assumed to have had little effect on yields.

A second trial of a subset of the F_1 hybrids was grown at NRPB for comparative purposes. These results are not described in detail.

3. RESULTS

Among the 36 F_1 hybrids, three are reconstituted tall genotypes, e.g. Hrht (♀) x Hrht (♂), Wrht (♀) x Wrht (♂). A comparison of these with the normal self-pollinated tall isogenic lines allows a test of any effects due either to the hybridising chemical itself, or to effects resulting from the lower fertility of the male-sterile females.

Nine are intravarietal F_1s, e.g. HRht1 x Hrht, WRht3 x Wrht. These are effectively isogenic lines with the Rht alleles present as heterozygotes, which can be compared directly with the homozygous rht and Rht parental isogenics in the same set.

The remaining 24 F_1s are all intervarietal hybrids, i.e. H x W, H x B or W x B, among which 12 genotypes are present as reciprocals, e.g. WRht1 (♀) x Hrht (♂) and HRht1 (♀) x Wrht (♂). No consistent differences were found between reciprocals and all the results have been meaned in the analyses described below. Comparisons of the intervarietal F_1s with the intravarietal hybrids, e.g. HxWRht1/rht with HRht1/rht and WRht1/rht, provides a direct measure of the heterotic effects due to the heterozygous background genotype, in the presence of, but not confounded with, the effects of identical alleles at the Rht loci.

3.1. Effects of homozygous and heterozygous dwarfing genes in isogenic backgrounds

Seven different genotypes, four homozygotes and three heterozygotes, differing only in the allelic status of the Rht loci, and, of course, any residual variation present after the backcrossing process by which the parental lines were produced, were available in each set to measure the effects of the genes on height and any pleiotropic effects on other characters.

The yield data for the three varietal series, with the Rht1 and Rht2 heterozygotes and homozygotes pooled, are shown in Figure 1. Clearly, on average, the Rht1 or Rht2 homozygotes have the highest yield potential. Nevertheless, this may, at least in part, be dependent on genetic background since in Bersee the Rht3/rht heterozygote is the superior genotype. This may be related to the fact that, of the three varietal backgrounds, Bersee is the tallest. This is so at all Rht allelic levels. The BRht3/rht heterozygote was 96 cm while the comparable W and H genotypes were 87 cm and 75 cm respectively.

The results for yield, height and ear yield components are shown, meaned over the three isogenic series, in Figure 2. The marked effects of the Rht alleles on grain number per ear and grain weight are closely correlated with the potency of the GA-insensitive gene combinations as measured by plant height reduction (Figure 2b).

The grain number per ear value for the Rht3/Rht3 genotypes, although high, is probably an artifactually low estimate. HRht3 had only 50 grains per ear, while Hrht had 36 and HRht3/rht had 56. Of the three varieties Huntsman is the shortest and HRht3 was the only line in which the ear did not fully emerge from the boot until well after anthesis, which will have had an adverse effect on the ear yield components of this genotype.

3.2. Effects of dwarfing genes on heterosis in F$_1$ hybrids

As mentioned above, the orthogonal nature of the F$_1$ hybrid combinations available allowed the effects of the dwarfing genes, at four allelic levels, i.e. rht/rht, Rht1/rht, Rht2/rht and Rht3/rht, to be compared in homozygous isogenic backgrounds and in heterozygous intervarietal backgrounds.

These two sets of data, for height, yield, ear yield components and harvest index are shown in Figure 3, pooled over the three intravarietal combinations, i.e. H x H, W x W and B x B, and the three intervarietal combinations, i.e. H x W, H x B and W x B.

The yield data in Figure 3a shows a clear interaction between the heterotic advantage in intervarietal F$_1$s over the intravarietal hybrids and the dwarfing gene allelic level. The tall intervarietals have only a 1.9% advantage, whereas the Rht3/rht intervarietals have an 8% yield advantage over comparable intravarietal F$_1$s.

This interaction is clearly due to a sustained effect of F$_1$ advantage in grain size of about 6% over all four height genotypes (Figure 3d), which operates on varying grain numbers per ear. The Rht3/rht F_1s derive most heterotic increase in yield because they have most grains per ear - 32% more, in fact, than the rht/rht hybrids (Figure 3c).

3.3. The source of heterosis

Ear density (data not shown) showed no differences between the intervarietal and the intravarietal F_1s, although there was a non-significant reduction of about 10% in the Rht3 genotypes relative to the semi-dwarf and tall lines. No heterotic differences were found for harvest index (Figure 3e), indicating that the effects of heterotic vigour, where present, were expressed throughout the whole plant and affected both straw and grain yields similarly. The general height increases in the intervarietal F_1s of about 5% (Figure 3b) are a reflection of this same effect.

The heterotic effects on grain weight were demonstrable in 10 out of the 12 intervarietal F_1s. The hybrid values are plotted between the intravarietal F_1s with identical dwarfing allele status in Figure 4. Only in two tall crosses do the F_1s have lower values than the higher parent.

3.4. The effect of the WL 84811 on F_1 yields

A comparison of the yields of the self-pollinated rht lines and reconstituted CHA produced intravarietal rht lines is shown in Table 1. No significant differences were detected for yield or the other parameters measured.

4. DISCUSSION

Among the hybrids studied here, in homozygous, intravarietal backgrounds the optimal genotype was plainly Rht2/rht, with a yield advantage of 4%, 5% and 13% over the Rht1/rht, Rht3/rht and rht/rht genotypes respectively. In heterozygous, intervarietal backgrounds Rht2/rht was still the highest yielding genotype (17.2% over the tall inbreds), but was closely followed by Rht3/rht (16.6%) which, in turn, showed more potential than Rht1/rht (12.2%) (Figure 3a). It is probable that this situation may vary with season and genetic background. Indeed, the highest yielding F_1 in this trial was Hrht x BRht3 at 1.3 kg m^{-2}.

Similarly, in the comparative trial at Nickerson RPB, the same genotype was again the highest yielding F_1 hybrid. In fact, it outyielded the four present-day varieties, included in the trial as checks, by 5%. This is in spite of the fact that these check varieties postdate Huntsman and Bersee by more than a decade.

Other considerations may favour the use of Rht3 in F_1 hybrids. Figure 2b shows that the Rht3/rht F_1s are some 12 cm shorter than comparable semi-dwarf inbreds. In the UK the higher yielding semi-dwarf varieties are already so tall that they can sucumb to lodging under high inputs. An additional height increase in intervarietal inbreds (Figure 3b), associated with heterotic increases in yield in Rht1/Rht1 or Rht2/Rht2 F_1 hybrids, could cause lodging to be a general and extensive problem for these genotypes.

There are, of course, other possible solutions besides the use of Rht3, which have not as yet been tested. These include the use of other non-GA-insensitive dwarfing genes [2] alongside Rht1 or Rht2, and the use of three Rht1 or Rht2 alleles in F_1s, i.e. Rht1/Rht1, Rht2/rht or Rht1/rht, Rht2/Rht2, obtained by crossing single gene semi-dwarfs with two gene dwarf lines.

The finding that increased grain size is the major source of F_1 heterosis for yield confirms both our previous results and those of the commercial F_1 hybrid programme at the Plant Breeding Institute. This is in spite of the fact that the three varietal combinations explored in the present experiment give relatively low levels of yield heterosis over the high parent (2-10%) compared with the 10-12% commonly found among the best Plant Breeding Institute hybrids [3].

The lack of obvious heterotic effects on other yield components indicates that breeders should actively seek ways of maximising the effect on grain weight. Increasing available grain sites by the use of Rht genes may be most applicable to breeders in winter wheat areas because the effect of the GA-insensitive genes on grain number may not be so marked in spring wheats [2]. The use of varieties characterised by high grain numbers, rather than those with large grains, might be expected to produce the same result.

5.   CONCLUSIONS

If F_1 hybrids emerge as a feasible commercial proposition then the use of Rht3 and tall parents should be strongly considered. Rht3/rht F_1s can provide high yield potential and an ideal straw strength phenotype for intensive, high yielding, high input agricultures such as Northern Europe. This would, of course, require the initiation of hybrid parent breeding programmes to produce both of the necessary tall and Rht3 dwarf

parental stocks, neither of which are primary targets in
present day commercial inbred programmes.

REFERENCES

[1] GALE, M.D., SALTER, A.M., CURTIS, F.C., ANGUS, W.J.
     (1988) The exploitation of the Tom Thumb dwarfing
     gene, Rht3, in F_1 hybrid wheats. In: Semi-dwarf
     cereal mutants and their use in cross-breeding
     III. Proc. Final FAO/IAEA Research Co-ordination
     Meeting, Dec. 1985, Rome, Italy. IAEA, Vienna,
     TECDOC-455: 57-68.

[2] GALE, M.D. and YOUSSEFIAN, S. (1985) Dwarfing genes
     in wheat. In: "Progress in Plant Breeding".
     G.E. Russell (ed.), Butterworths, London, 1-35.

[3] BINGHAM, J. (1986). Adaptation of new techniques in
     wheat breeding. Plant Breeding Smyposium DSIR,
     Christchurch, NZ. Williams T.A. and Writt, G.S.
     (eds), Agronomy Soc. NZ Special Publication No. 5:
     97-103.

TABLE I Yields (g.m.$^{-2}$) of three varieties, sown from grain
harvested from conventional plots and from CHA
'hybrid' plots

Source of grain	Hrht	Wrht	Brht
Self pollination	1187	992	1055
CHA self hybridisation	1205	968	1044

LSD 5% = 80.6 g

FIGURE 1.   The effects of allelic differences at the <u>Rht</u> loci
on grain yield in three near-isogenic series.

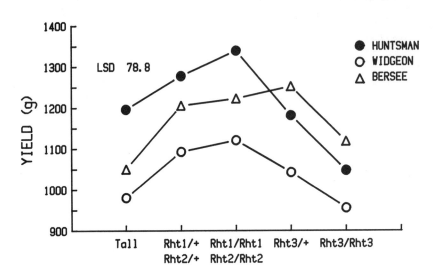

HUNTSMAN, WIDGEON & BERSEE SERIES

Note.   The <u>Rht1/rht</u> and <u>Rht2/rht</u>, and the <u>Rht1/Rht1</u> and
<u>Rht2/Rht2</u> means have been pooled within each series.

FIGURE 2. Grain yield, plant height, and ear yield components
of seven <u>Rht</u> allele combinations meaned over the
Huntsman, Widgeon and Bersee near-isogenic series.

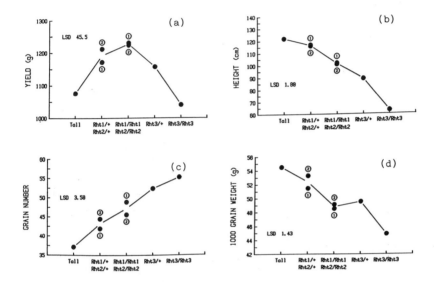

Note. The <u>Rht1</u> and <u>Rht2</u> heterozygotes and homozygotes are
indicated by (1) and (2).

FIGURE 3. Comparisons of yield, height, ear yield components
and harvest index for rht/rht, Rht1/rht, Rht2/rht
and Rht3/rht genotypes in homozygous (intravarietal)
and heterozygous (intervarietal) genetic
backgrounds.

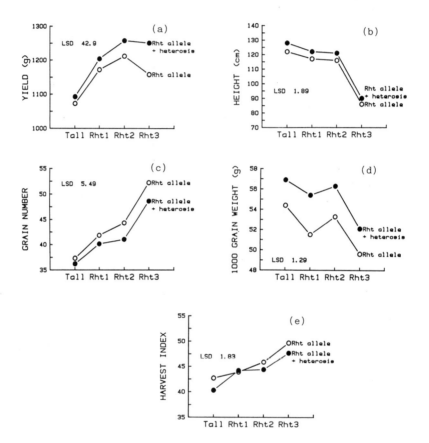

Note. The intravarietal points, O, are the means of 12 values
(3 isogenic F₁s x 4 blocks), while the intervarietal
points, ●, are the means of 24 values (2 reciprocals x 3
intervarietal F₁s x 4 blocks). The LSD 5% value applies
to differences between the two sets of points.

FIGURE 4. Grain weights of 12 F_1 hybrid combinations, in three intervarietal combinations at four, rht/rht, Rht1/rht, Rht2/rht and Rht3/rht, dwarfing allele levels.

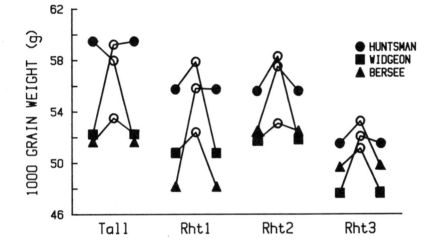

Note. For each cross the mean of the reciprocal intervarietal F_1s (e.g. HRht1 (♀) x Wrht (♂) and WRht1 (♀) x Hrht (♂) is plotted between the isogenic values for the same Rht allele combination (e.g. HRht1 x Hrht and WRht1 x Wrht). The intervarietal points are the means of 8 values, the intravarietal points are the means of 4 values. The LSD 5% for comparisons between the intra- and intervarietal points is 2.12 g per 1000 grains.

A SUMMARY ON THE CURRENT STATUS OF HYBRID RICE IMPROVEMENT BY USING INDUCED MUTATIONS AND IN VITRO TECHNIQUES

GAO MINGWEI

Zhejiang Agricultural University
Hangzhou
China

M. Maluszynski (ed.), Current Options for Cereal Improvement, 63–70.
© 1989 by Kluwer Academic Publishers.

ABSTRACT

According to surveys made in recent years, more than 30 sources of cytoplasmic male sterility can be identified in rice, among which only six are being commercially used in China. Although hybrid rice of the WA cyto-sterility system has been widely cultivated in China, laborious procedures and low yield in hybrid seed production, seed contamination in the process of multiplication and hybrid production or other drawbacks call for further research efforts. One of the most effective ways to solve these problems is the application of mutation induction and in vitro techniques. New restorer lines were already isolated from the irradiated progenies. The hybrids with new restorer lines outyield the locally prevailing hybrid combination, mature earlier, and have better grain quality.

At least 100 varieties or promising lines including rice hybrids were so far developed in China by using anther culture. The yield of anther-derived progenies of hybrids was usually comparable to the initial F_1 hybrids, but much higher than the parental restorer lines.

INTRODUCTION

China was the first country to use cytoplasmic male sterility to develop hybrid rice for commercial use in 1973. In 1986 more than 8 million hectares of hybrid rice were planted in China, which is one fourth of the total rice area and produces one third of the total rice in the country. Hybrids usually outyield the leading commercial varieties by 20-30%, giving an average yield advantage of 1 to 1.5 t/ha, because of their better morphological traits, higher physiological efficiency, better resistance to major diseases and insects, and wide adaptability to various agro-ecological stresses.

IMPROVEMENT OF HYBRID RICE

A. Mutation techniques
Almost all of the cultivated F_1 rice hybrids in China are developed from cytoplasmic male sterile and restorer lines. According to surveys made in recent years, more than 30 sources of cytoplasmic male sterility in rice can be identified, among which only six are being commercially used (Table 1).

Wild rice with aborted pollen (WA) cytosterility system is the most popular one in use to develop male sterile lines (MS line) in China. The main technique available for developing stable MS lines is substitution backcrossing of the genome of one species into alien cytoplasm of another. Sufficient backcrosses are required to eliminate all nuclear genes derived from the cytoplasm donor species. A number of studies have shown that using interspecies crosses, such as the cross of wild rice (O. perennis, O. sativa, f. spontanea, etc.) with indica varieties, and intersubspecies crosses, such as the cross between indica and japonica rice, usually produce stable MS lines.

Most restorer lines in the WA cytosterility system possess two dominant R genes which are capable of restoring the fertility of MS lines [1]. There are several ways to develop R lines, among the most reliable of which is to screen them from the existing R sources by using the testcross method. Results revealed that most of the R lines of the WA cytosterility system were derived from late maturing indicas such as IR lines from IRRI and many varieties from Southeast Asia. They originated from the habitat at lower altitude and were closely related to wild rice.

Although hybrid rice of the WA cyto-sterility system has been widely cultivated in China, there are still problems. First, most of the currently cultivated hybrids have a longer growth duration (125-140 days) than conventional varieties and thus cannot be grown during the first cropping season (March/April to June/July) in the vast area in south and east China. This season is short because of the low temperature prevailing before March/April. Second, the lack of multiple disease and insect resistance and the loss of rice blast resistance in the current hybrids has prevented their wider spread and their yield heterosis from being fully exploited. Third, poor grain quality, unacceptable to the world market. In addition, laborious procedures and low yield in hybrid seed production, seed contamination in the process of multiplication and hybrid production, and other drawbacks call for further research efforts.

One of the most effective ways to solve these problems is the application of mutation induction and in vitro biotechniques. Breeders at Sichuan Institute of Nuclear Technology, China, irradiated the restorer line Taiyin 1 with 60 KR gamma rays in the 1970's with the objective of obtaining earlier and shorter mutant restorers. The result was very exciting. A new restorer line Fuhui 06 was finally isolated from the progenies of the treated population in 1980, turning out to be 20 days earlier and 10 cm shorter than the original restorer. The hybrid between Gam type MS line and restorer Fuhui 06 outyielded the locally prevailing hybrid combination Shan You 2 by 4%, maturing 7 days earlier and possessing a content of crude protein as high as 15% [2].

Another example was the development of the restorer line Fu 36-2, a mutant derived from the mother variety IR 36 by breeders in Zhejiang Province, east China. IR 36 was introduced to China from IRRI in the 1970's and has proven to be a good restorer of WA cytosterility through repeated test crosses, but it showed a high sensitivity to temperature, thus narrowing the adaptability of the F_1 hybrid to a limited agro-ecological environment. Besides, the hybrids also have the defects of unstable fertility and unsatisfactory grain quality. With a view to overcoming these drawbacks, dry seeds of IR 36 were irradiated with 30 KR gamma rays in autumn, 1977. About 60 individuals were isolated from nearly 10 thousand M_2 plants. Sixteen of these were selected in M_3 generation and finally, 3 mutant families were screened in M_4 for further evaluation. From M_5 generation on, the selected lines were stable in mutated characters, then test crossed to MS line to observe the fertility restoration. Experimental results showed that Fu 36-2 retained the basic agronomic traits and restoration ability remained unchanged, as IR 36, but has overcome those parental defects [3]. Consequently, the cross combination between the MS line Zhen Shan 97A and the new R line Fu 36-2 presents a hybrid yield parallel to the currently cultivated hybrid Shan You 6, but with better grain quality, wider adaptability as well as enhanced resistance to blast. Now, Fu 36-2 is being used as a new restorer line in the province.

Mutation induction also helped improve the grain quality of rice hybrids. Most of the MS lines and R lines in China do not possess good grain quality. Consequently, the quality of the hybrids is also poor. Since the kernels harvested from the F_1 hybrids are F_2 seeds which would segregate for grain characteristics, it is imperative that traits for grain quality should be improved in both MS lines and R lines. Currently, a number of R lines with good quality (which were developed by conventional breeding methods) are already available in China, but it is not an easy job to obtain a MS line with good grain quality by using the same method. A few maintainer lines possessing all the characters such as high receptibility of alien pollens, high combining ability and acceptable grain quality can be found among the existing rice varieties. An alternate resort is mutation techniques. Breeders in Zhejiang Province, treated the indica cultivar Hon 410 with physical mutagens in 1977 to induce mutations for good grain quality and have obtained very encouraging results. A new mutant variety, Hontou 31, was developed in 1982 and officially released in 1985. Hontou 31 possesses good grain quality and desirable flowering habit, and was immediately crossed with WA cytosterile line Zhen Shan 97A in early 1982. Afterwards, a new MS line Hontou 31A was formed in 1984 through successive backcrossing to MS line Zhen Shan 97A for six generations [4]. Now, Hontou 31A is being used in hybrid seed production, representing a breakthrough in improvement of hybrid grain quality.

Breeders in China also treated other elite restorer lines such as: IR 24, T-64, Minghui-63, etc., with gamma rays and chemical mutagens in an attempt to improve agronomic traits and rice blast resistance of the initial lines. Research work has been in progress and a lot of selected lines are now under field evaluation.

B. In vitro cultures

In recent years, in vitro techniques have been successfully incorporated into hybrid rice breeding in China. A number of studies revealed that in vitro culture of young inflorescence and immature embryos of hybrids showed higher frequency of callus induction and shoot differentiation, greater growth rate and fewer albino regenerants compared to that of conventional varieties. Indica hybrids tend to respond better to somatic tissue culture than japonica hybrids. The regenerated plants were found morphologically identical to the F_1 hybrids in plant-type, head-type, and other traits of economical importance. It seems that the regenerants would be able to maintain the same level of heterosis as F_1 hybrids. Occasionally, chromosome aberrations including changes of chromosome number (polyploidy and aneuploidy) and abnormal chromosome behaviour have been observed in a few regenerated plants, which were derived from immature embryos. Most of the mutated characters in the second generations occurring in the initial somatic tissue have become homozygous through self-pollination. On the other hand, a few variants still segregated for culm height and leaf-type even in F_4 generation.

With regard to doubled haploids in rice, China has so far developed at least 100 varieties and promising lines including rice hybrids by using anther culture. The yield of anther-derived progenies of hybrid was usually comparable to the initial F_1 hybrids, but much higher than the parental restorer lines. This could be considered a novel way to fix the high yielding composition of genes [5].

Besides, a number of doubled haploids derived from anther culture of male sterile lines have also been

developed since 1980 and were used as pure female parents in
hybrid production. Most of the DH male sterile lines proved
identical to the initial parent in morphological and
agronomic traits. The fertility of the male sterile doubled
haploids could be restored by the same pollen donor lines.
The hybrids coming from the crosses of DH male sterile lines
with the same restorer line, in most cases, exhibited
greater heterosis, outyielding the counterpart hybrids by
17% on the average and by a maximum of 29% [6]. Obviously,
the yield advantage resulted solely from the function of
purification of the male sterile line through anther
culture.

C. Environmentally induced sterility

In recent years an environmentally induced sterility
system has been established in China in the latter part of
the 1970's. A male sterile line grown under long day-length
conditions was found to be fully fertile when it was grown
in a short day-length season. Studies showed that the
photo-period induced male sterility is controlled by a pair
of recessive genes and is irrelevant to cytoplasmic factors
[7]. The genes have been transferred to many other
cultivars to form new male sterile lines. The use of this
particular sterility system will provide more opportunities
for selecting superior combinations and lead to cost
reduction of hybrid seeds.

Table 1:   Cytoplasmic sources widely used to induce male
              sterility for commercial hybrid rice

Cytoplasm	Nuclear Source	Type of Sterility
I. <u>O</u>. <u>sativa</u> <u>f</u>. <u>spontanea</u> <u>L</u>.	<u>indica</u> rice	
(1) Wild rice with aborted pollen (WA)	Zhen Shan 97, V20 and many others	Sporophytic,shrunken pollen unstained in I-KI solution
(2) Dwarf wild rice	Xie-Ging Zao and many others	"
II. <u>O</u>. <u>indica</u>	<u>japonica</u> rice	
(1) Chinsurah Boro II	Tai Chung 65	Gametophytic, rounded pollen, only a few stained slightly
(2) Lead	Tengban 5 and number of others	"
(3) Eshan-Dabaigu	Hong-Mao-Ying <u>indica</u> rice	"
(4) Gambiaca	Zhao-Ying 1	Sporophytic, shrunken pollen - unstained

REFERENCES

[1] GAO MINGWEI, (1981)   A preliminary analysis of the genotypes of hybrid indica rice with wild rice cytoplasm.   Acta Genetica Sinica, 8 (1): 66-74.

[2] LEE XIAO, (1983) A brief comment on the mutant restorer line "Fuhui 06". Sichuan Agric. Sci. & Tech. 6: 36.

[3] MAO XINYU, (1986)   A new restorer line Fu 36-2 developed by using gamma irradiation in indica rice. (in press)

[4] YEH ABAO, (1986)   Hontou 31, a rice mutant with early maturity and superior grain quality.   Zhejiang Agri. Sci. 2: 59-61.

[5] GE MEIFENG, TAN CHANGLE and BAI HESHENG, (1986) Anther culture of hybrid rice and male sterile line of Oryza sativa subsp. indica and its application to rice improvement. Jiangsu Journal of Agri. Sci. 2 (1): 1-6.

[6] GE MEIFENG, TANG CHANGLE, BAI HESHENG and YANG QIANGPING, (1985)   Anther culture and its application to purification and rejuvenation of male-sterile line of wild abortive rice system. Jiangsu Journal of Agric. Sci. 1 (1): 25-32.

[7] SHI MINHSONG, (1985)   The discovery and study of the photo-sensitive recessive male-sterile line in rice (Oryza sativa L. subsp. japonica). Scientia Agricultura Sinica, 2: 44-48.

PRODUCTION OF HAPLOIDS IN CEREALS

K.J. KASHA

Crop Science Department
University of Guelph
Guelph, Ontario, Canada

M. Maluszynski (ed.), Current Options for Cereal Improvement, 71–80.
© *1989 by Kluwer Academic Publishers.*

Abstract:

There are a number of methods for producing haploid plants
in cereal crops with anther (microspore) culture being the most
universal. Others are the Bulbosum Method, ovule culture and
genetic or cytoplasmic factors. The Bulbosum Method is
restricted so far to barley and wheat and is, to-date, the
proven method for haploid barley production. Ovule culture has
worked in barley, rice and wheat while genetic factors are
known in barley, maize and wheat. From the perspective of
using mutagens, isolated microspore culture might be preferred
but has not been perfected as yet for cereals. Our efforts at
Guelph on developing haploid systems have shifted from the
Bulbosum Method to the development of anther and isolated
microspore culture for wheat and barley in the past four years.
We have developed a defined media, BAC1, which allows us to
more critically examine the various media requirements. A
liquid media with Ficoll and with replenishment every 10 days
has proven to be effective. This permits the initial use of
high auxin and sucrose and the subsequent shift to low sucrose
and other hormones. High levels of inositol are important and
permit the use of lower auxin concentrations. Various problems
require further research; strong genotype variation in
response, albinism, regeneration frequencies and the cause and
extent of gametoclonal variation. Cytological studies of
barley anther culture provide evidence that the chromosomal
variation occurs in the early stages of anther culture.

1.0 Introduction:

Haploid cells with only one set of chromosomes could be an
ideal unit for the induction of mutations in crop plants. The
limitation is to be able to regenerate plants from haploid
cells. Ideally, the haploid cells could be the numerous
immature microspores produced following meiosis. Therefore,
methods of producing plants from such gametes have been, and
are, of great interest. Alternatively, once haploid plants
have been produced from a desired line, cell cultures can be
established to produce large populations of haploid cells for
mutation and selection. Both cell and microspore cultures may
also lead to variability(somaclonal or gametoclonal,
respectively) and serve as an alternate approach to obtaining
desired variants.

In this overview on haploid production in cereals I will
generalize on procedures and problems but utilize examples from

our own work on barley and wheat to illustrate points. Any
system selected to produce haploids should be able to produce a
large random sample of haploid plants from any given genotype.
The plants should be genetically stable and their production
should be highly efficient.

2.0 Methods of producing haploids:

Haploids can be produced in cereals by a number of
procedures as summarized in Table I. The most universal and
potentially the most efficient method is anther or isolated
microspore culture. The other procedures are restricted to the
maximum of one haploid per floret, compared to the thousands of
microspores. However, the Bulbosum Method, resulting from the
cross of barley or wheat with pollen from the wild species
Hordeum bulbosum, is the proven method for haploid production
in barley at the present time.

It may be of interest to this working group that our
research objective that led to the discovery of the Bulbosum
Method was to develop material for hybrid barley production.
Our plan was to produce interspecific hybrids between barley
and Hordeum bulbosum and irradiate these hybrids in an attempt
to induce the transfer of cross-pollination traits into barley
chromosomes. Instead, we observed the phenomenom of chromosome
elimination in the hybrids which, combined with embryo culture,
led to the production of barley haploids[1]. Species crosses
with the parental genomes in the ratio of 1:1 generally
resulted in the elimination of the H. bulbosum chromosomes,
whereas, a genome ratio of 2 bulbosum : 1 barley tended to
produce triploid hybrids that underwent very limited
elimination of chromosomes. Ho and Kasha[2] later demonstrated
that this balance phenomenon was restricted to factors on
chromosomes 2 and 3 of barley and their balance with H.
bulbosum chromosomes.

Barclay[3] demonstrated that the pollination of cv.
Chinese Spring wheat with tetraploid H. bulbosum also led to
chromosome elimination and the production of haploids in wheat.
However, in wheat, the method was restricted to only a few
lines which carry the recessive alleles of the crossability
genes Kr1 and Kr2[4,5]. Because of this crossability
limitation, we decided to pursue anther culture as a system for
haploid production in wheat. At that time, there were reports
of some success[6,7] with anther culture in barley and we
therefore examined these procedures in both crops as well as to
a limited extent in triticale.

The other procedures mentioned in Table I, while they can
produce some haploids, have not been perfected to the stage of
their consideration for breeding and I shall not discuss them
further here.

3.0   Factors influencing anther culture success:

These factors may be described under four general areas:
1. The condition of the donor plants
2. Environmental conditions during plant growth and at the time of anther culture.
3. Culture media components
4. The donor plant genotype.
Because of these factors and their interactions, it is usually difficult to repeat the results reported from other labs. Some of these factors appear to need adjustment for any specific location.

The vigour and health of the donor plants is critical to success in microspore culture. Relative to environmental conditions, most researchers tend to grow their plants at relatively low temperatures and often use a cold pretreatment at the time when microspores reach the correct stage for culture. Ouyang et al.[8] demonstrated that a high culture temperature of 30-32^{0}C was important for anther culture in wheat. We have obtained similar results and also examined the development of multicellular structures (Table II). Culture temperature is also important in barley and we have observed the maximum effect at 28^{0}C. Our results were obtained without using cold pretreatments of spikes or anthers. We have not been able to show a beneficial effect of cold pretreatment in wheat[9] nor in barley(unpublished). Perhaps this could be due to the relatively high temperatures(16-21^{0}C) at which our plants are growm in the growth room, but we have not verified this assumption.

Culture media factors are numerous and are often associated with genotype effects within a species. The search for a universal culture media upon which most genotypes will respond continues today. Our attempts to follow the barley anther culture procedures of Kao[6] and Xu and Sunderland[7] were unsuccessful and therefore, we embarked upon experiments to more precisely examine the media components for barley and wheat. For this purpose, we required a medium that could be defined (not including potato extract, or coconut milk, etc.). Based on results in the literature, Marsolais[10,11] put together a defined medium called BAC1 (barley anther culture medium 1). This was a liquid medium with Ficoll and therefore could be modified or replenished at any time during culture. With this medium, callus induction appeared to be highest in barley using 9% sucrose and 8mg/L 2,4-D and replenishment every 10 days with the same medium except using 3% sucrose and 1mg/L IAA in place of the 2,4-D[11]. The same medium worked well for wheat anther culture although Kelly[12] observed that 6% sucrose and 1.75% glucose was better in the induction phase.

With the BAC1 medium, the myoinositol concentration proved to be a critical factor. When the concentration was increased from the usual 100mg/L to 1000mg/L the number of calli produced

more than doubled. Subsequent experiments showed the optimum
level to be in the range of 2.0 to 3.5g/L(Table III). At these
high inositol levels the type and concentration of auxin could
also be changed in the BAC1 medium(Table IV). Lower levels of
2,4-D were effective and IAA could be used in place of 2,4-D,
thus reducing concerns about mutation and albinism possibly
associated with high levels of 2,4-D.

Hunter[13] and Shannon et al.[14] demonstrated that the
orientation of the anthers on edge on solid culture media had a
marked effect on response. Using the BAC1 medium solidified
with agar we have not found any effect of anther orientation
with either barley or wheat relative to the numbers of anthers
responding or of calli produced. Hunter suggested that the
microspores remained in the top locule of the anther when on
edge but dehisced into the medium when lying flat and the
latter did not develop further. This may be related to the
anther factor in barley described by Kohler and Wenzel[15].
If the microspores are removed from the anthers it is likely
that they will require conditioned media or the identification
and addition of the anther factor(s). It is possible that the
media density may also influence anther response as illustrated
in Table V but this requires further study relative to anther
dehiscense.

In liquid culture media the concentration of Ficoll is
also important and can interact with the genotype(Table VI).
The first three genotypes in Table VI are barley while the
fourth is wheat. In general, it appears that the higher
concentrations of Ficoll are better and 200 to 300g/L are now
recommended. However, Ficoll is an expensive chemical and
cheaper alternatives such as high molecular weight dextran or
soluble starch may be as effective(Table VII). Kelly[12]
observed that 80g/L dextran in the liquid media was as
effective as 100g/L of Ficoll on two cultivars of wheat.

4.0 Genotype Response:

Prior to the change to the higher inositol concentration
in the BAC1 medium, we had conducted small surveys of anther
culture response in winter wheat genotypes, using both solid
and liquid media. On medium solidified with Bacto-Agar, all 19
genotypes surveyed responded with the range being 0.4 to 19%
and the mean 6.6% of the anthers plated. On liquid media with
100g/L Ficoll, the range was 2.0 to 35.2% with a mean of 15.1%
anther response among the nine genotypes surveyed. Marsolais
et al.[16], using 2.0g/L myoinositol and 100g/L Ficoll
attempted to produce haploids from five winter wheat F1
hybrids. Three of the hybrids had one high responding parent
and averaged about 15% anther response. The two hybrids
without a high responding parent had 2.1 and 3.7% anther
response and resulted in only 9 green plants out of the 450
produced. The 450 green plants required a total of 121
technical hours of work so that, on average, over a 100 green

plants were produced per week of work. Such an efficiency is worthy of consideration in a breeding program in spite of the wide genotype variation.

5.0 Induced Variation:

We have produced a number of spontaneously doubled haploids from the wheat cultivar Sinton and compared their performance in replicated field trials relative to the parental cultivar. Twenty-four of the 30 doubled haploid lines evaluated were significantly different from the parent for at least one of five agronomic traits studied. Studies on the heritability of some of these trait variations are continuing. The extent of the variation observed was quite surprising however, with the better embryoid formation and frequencies now available[16] it is unlikely that the variation would be as extensive.

Chen et al.[17,18] studied the process of cell development and chromosome variation during the first 10 days of barley anther culture. After two days in culture, over 99% of the microspores had undergone a mitotic division. About 12% of the microspores had an abnormal division in that the typical generative cell was not formed. Most often these variants had two vegetative-type nuclei with no cell wall present between them. Through a combination of light and electron microscopy they concluded that nuclear fusion was occurring in the early stages of culture, leading to the production of diploid and tetraploid progeny. In addition, a few apparent instances of C-mitosis were observed which could lead to the formation of diploid and aneuploid cells. Nearly 50% of the green plants produced are homozygous diploid plants in barley and wheat while about 10% in wheat are aneuploid.

6.0 Conclusions:

The efficiency of anther culture in most cereals has reached the stage where it can be used in breeding programs to some extent and many new cultivars produced by this system have now reached the market. For the purposes of mutation research and breeding, it would be desirable to use isolated microspore culture so that the mutagens could be applied more uniformly and the results more readily observed and followed. To-date, there has been only one report of isolated microspore culture success in barley[19] and other labs have had difficulty in repeating that procedure. We have worked on isolated microspore culture for the past couple of years. While we are able to isolate the microspores and keep them in a viable and apparently healthy condition for over 10 days, very few exhibit nuclear and cell division and none of the responding ones have regenerated into a plant. Up until now, we have attempted to work with defined media and without preculture in order to try to identify the critical media factors. Recent success has been reported in wheat[20] with shed pollen and conditioned media, again illustrating the importance of a factor in the anther or ovales. It is likely that we will have to resort to conditioned media to obtain results at this time. Isolated microspore culture in cereals will be our main thrust in haploidy over the next period of time.

7.0 Acknowledgements.

Many Research Associates, Graduate Students and Technicians contributed to the results summarized in this paper and I wish to acknowledge the following: Dr. A.A. Marsolais, Dr. W.G. Wheatley, Dr. A. Ziauddin, S. Kelly, J. Lettre, K. Glover, E. Cober and R. Oro. Research funding and/or facilities from the Natural Sciences and Engineering Research Council of Canada and the Ontario Ministry of Agriculture and Food are most gratefully acknowledged.

8.0   REFERENCES

[1] KASHA,K.J., KAO,K.N., High frequency haploid production in barley(Hordeum vulgare),Nature(London)225(1970)874.
[2] HO,K.M.,KASHA,K.J., Genetic control of chromosome elimination during haploid formation in barley,Genetics81(1975)263
[3] BARCLAY,I.R.,High frequencies of haploid production in wheat (Triticum aestivum) by chromosome elimination, Nature(London) 256(1975)410.
[4] FALK,D.E.,KASHA,K.J., Genetic studies of the crossability of hexaploid wheat with rye and Hordeum bulbosum, Theor. Appl. Genet. 64(1983)303.
[5] SNAPE,J.W., et al. The crossabilities of wheat varieties with Hordeum bulbosum,Heredity 42(1979)291.
[6] KAO,K.N., Plant formation from barley anther cultures with Ficoll media, Z. Pflanzenphysiol. 103(1981)437.
[7] XU,Z.H.,HUANG,B.,SUNDERLAND,N., Culture of barley anthers in conditioned media, J. Exp. Bot.32(1981)767.
[8] OUYANG,J.W.,ZHOU,S.M.,JAI,S.E., The response of anthers to culture temperature in Triticum aestivum, Theor. Appl. Genet. 66 (1983)101.
[9] MARSOLAIS,A.A.,SEGUIN-SWARTZ,G.,KASHA,K.J., The influence of anther cold pretreatments and donor plant genotypes on in vitro(Triticum aestivum), Plant Cell Tissue Organ Culture 3(1984)69.
[10] MARSOLAIS,A.A., Callus induction from barley microspores, Ph D thesis,Univ. Guelph(1985).
[11] MARSOLAIS,A.A., KASHA,K.J., Callus induction from barley microspores. The role of sucrose and auxin in a barley anther culture medium, Can. J. Bot.63(1985)2209.
[12] KELLY,S.A., The effect of sucrose ,dextran and glucose on anther culture of winter wheat(Triticum aestivum L.), M Sc. Thesis, Univ. Guelph(1986)103pp.
[13] HUNTER,C.P., The effect of anther orientation on the production of microspore-derived embryoids and plants of Hordeum vulgare cv. Sabarlis, Plant Cell Reports 4(1985)267.
[14] SHANNON,P.R.M., et al., Effect of anther orientation on microspore-callus production in barley(Hordeum vulgare L.), Plant Cell Tissue Organ Culture 4(1985)271.

[15] KOHLER,F.,WENZEL,G., Regeneration of isolated barley
     microspores in conditioned media and trials to
     characterize the responsible factor, J. Plant Physiol.
     121(1985)181.
[16] MARSOLAIS,A.A.,WHEATLEY,W.G.,KASHA,K.J., Progress in wheat
     and barley haploid induction using anther culture. DSIR
     Plant Breeding Symp. (1986)340.
[17] CHEN,C.C.,KASHA,K.J.,MARSOLAIS,A., Segmentation patterns
     and mechanisms of genome multiplication in cultured
     microspores of barley, Can. J. Genet. Cytol.,26(1984)475.
[18] CHEN,C.C., et al., Ultrastructure of young androgenic
     microspores in barley anther culture, Can.J. Genet.
     Cytol.,26(1984)484.
[19] WEI, Z.M., KYO, M., HARADA, H., Callus formation and
     plant regeneration through direct culture of isolated
     pollen of Hordeum vulgare cv. Sabarlis, Theor. Appl.
     Genet.,72(1986)252.
[20] DATTA, S.K., WENZEL, G., Isolated microspore derived plant
     formation via embryogenesis in Triticum aestivum, L.,
     Plant Science 48(1987)49.

Table I. Methods succeeding in producing haploids in cereals

Procedure	Barley,	Wheat,	Oats,	Triticale,	Rice
Anther culture	x	x	x	x	x
Bulbosum method	x	x			
Ovule culture	x	x			x
Microspore culture	x				
Hap gene	x				
Alien cytoplasm		x			

Table II.   Influence of culture temperature on anther
             culture response (A, cv. Sinton) and microspore
             development (B, cv. Pitic) of wheat.

	22	24	26	Temperature 28	30	32
A. response/100 anthers						
i. anthers	0.5	1.0	1.5	1.0	12.1	11.6
ii. embryoids	0.5	2.0	4.0	1.5	22.2	32.3
iii. green plants	0	0	0	0	1.0	2.0
B. #nuclei @ 10 days						
≥ 4	0	2	1	18	30	50
≥ 8	0	1	1	12	12	33
≥ 15	0	0	0	2	1	13
embryoids	0	0	0	1	3	14
Total scored	2575	2413	1856	1700	2405	1338

Table III.   Effect of inositol levels on haploid induction in anthers of barley cv. Klages using BAC1 Liquid media.

Inositol (mg/l)	% Anther response[a]	Calli per 100 anthers
1000	21.6 b	173.7 b
1500	23.0 b	152.1 b
2000	37.5 a	415.5 a
2500	31.1 ab	278.8 ab
3000	30.3 ab	247.0 ab
3500	37.8 a	444.0 a

[a]288 anthers plated per inositol concentration. Significant differences within a column are indicated by letters.

Table IV.   Effect of 4 levels of 2,4-D and IAA on haploid induction in barley anthers of cv. Klages using BAC1 media with 2.0 g/L myoinositol

Auxin conc. mg/l		% Anther response	Calli per 100 anthers
2,4-D	1	33.5	296.0
	2	35.3	299.2
	4	33.1	274.0
	8	21.5	147.4
IAA	1	21.8	263.8
	2	29.0	243.0
	4	32.7	320.5
	8	32.1	303.6

198 anthers were cultured at each concentration. No significant differences were observed.

Table V.   Effect of agarose concentration on anther response of 2 wheat genotypes.

Agarose conc.	% anther response	% E-type calli	% Total calli
0.45%	20.0	31.7	50.9
0.90%	13.4	19.4	33.4
0.45 VS 0.9%	**	*	*

**, * significant at 0.01 and 0.05% levels.

Table VI.   Means and significant differences of Ficoll
             concentrations and genotypes compared to anther
             response and productivity.

Genotype	Ficoll conc.(g.1⁻¹)	No. anthers plated	% anther response	calli per 100 anthers
Conquest	100	483	2.6 e	8.2 d
	200	483	1.5 f	3.2 e
Bruce	100	499	5.5 e	23.3 d
	200	495	12.0 c	40.9 c
Klages	100	418	35.2 b	356.7 b
	200	418	51.4 a	918.6 a
OAC 82-17	100	549	16.8 c	42.8 c
	200	534	7.7 d	15.0 d

Means within each column followed by the same letter are not
significantly different at the 0.05 level of probability
according to Duncan's Multiple Range test.

Table VII.   Evaluation of replacements for Ficoll in liquid
             BAC1 media, using anther culture of barley
             cultivar Klages.

Treatment	% Anther Response	Calli/100 Anthers
100 g/L Ficoll 400	37.0 a	272 a
100 g/L Dextran 249	36.6 a	241 a
100 g/L Soluble starch	31.5 a	241 a

Values are means and letters within columns indicate no
significant differences.

SOMACLONAL SELECTION OF PHYSIOLOGICAL MUTANTS THROUGH PLANT CELL CULTURE

T. KINOSHITA, K. MORI AND T. MIKAMI

Plant Breeding Institute
Faculty of Agriculture
Hokkaido University
Sapporo 060, Japan

M. Maluszynski (ed.), Current Options for Cereal Improvement, 81–96.
© *1989 by Kluwer Academic Publishers.*

Abstract

First, streptomycin resistance in rice callus clones was examined. Callus tissues were initiated from seed, radicle and anther cultures of rice (*Oryza sativa* L.) in order to study the effect of streptomycin on callus growth. Our results showed that the addition of 250 μg/ml or more streptomycin to the culture medium caused a significant inhibition of callus proliferation. The degrees of inhibition depended upon the genotype, the drug concentration and the tissue source of callus. Selection of resistant cell lines began with seed and immature embryo cultures grown on various levels of streptomycin. The fastest growing sectors of callus clones were isolated after 7 or 12 subcultures. Some of these clones exhibited a significant increase of resistance index when compared with unselected starting material. After 5 or 6 selection cycles, 79 plantlets were regenerated from resistant callus, but none grew to maturity because all were albino.

In the second experiment for somaclonal selection on salt tolerance, it was found that the callus tolerance is intensified by the repeated selections with the medium containing NaCl . Genotypic difference was also recognized among the calli from the three cultivars, Shiokari, Norin 8 and Taichung 65.

I.   Streptomycin resistance in rice callus clones

INTRODUCTION

Much effort has been directed towards obtaining a range of physiological and biochemical mutants of

higher plants by means of *in vitro* culture. One of
the most studied and well characterized examples is
cytoplasmically inherited resistance to streptomycin
in *Nicotiana tabacum* (Maliga et al., 1973, 1975; Umiel,
1979). More recently, Maliga (1981) has also indicated that streptomycin resistance in a cell line
isolated from haploid *N. sylvestris* is the result of a
recessive nuclear mutation.

Antibiotic resistant mutations could be useful
in an extensive analysis of the contribution of chloroplast and nuclear genes to ribosomes, as described
in the unicellular green alga, *Chlamydomonas* (Kirk &
Tilney-Basset, 1978). In addition, resistant phenotypes may be useful in somatic hybridization both for
the selection of fusion products and as a chloroplast
marker (Galun & Aviv, 1983). Therefore, we have initiated research to evaluate streptomycin for potential use in a selection system of rice (*Oryza sativa*
L.) tissue cultures. This communication reports the
inhibitory effect of streptomycin on the growth of
rice callus cultures. A preliminary result from selection of callus clones resistant to this compound
is also presented.

MATERIALS AND METHODS

A total of 111 rice genotypes including *japonica*
and *indica* type varieties and their hybrids were used
to isolate callus tissues from various organs. Primary calli were induced from dehusked seeds, immature
embryos (isolated 10 days after anthesis), anther
cultures, and from radicle sections (ca. 10 mm long)
of 8-day old seedlings grown in *in vitro*. Explants
were sterilized in 70% ethanol (1 min) and in 2% sodium hypochloride (30 min), followed by three washes
in sterile water. Anther cultures were started and
maintained on N_6 medium (solid) supplemented with

2 mg/1 2,4-D (Chu et al., 1975). With all other
explants, Murashige & Skoog (1962) solid medium (MS)
with 2 mg/1 2,4-D was used for tissue cultures and
assays.

Selection for streptomycin resistant cultures
was initiated by transferring a total of 32 callus
pieces (derived from seed and/or immature embryo
cultures) weighing approximately 50 mg onto MS medium containing 0, 250, 500 and 1000 µg/ml of streptomycin. In each subsequent selection cycle (one
subculture transfer), the fastest growing sectors of
callus were divided into 50 mg pieces and placed on
the fresh medium with various drug concentrations.
Transfers were made at 5- to 8-week intervals. Plant
regeneration was accomplished in MS medium containing 10 mg/1 kinetin. Cultures were incubated in
light at 26°C, but callus induction was done in
darkness.

RESULTS

Callus induction. White or light-yellow calli were readily obtained for all genotypes examined by culturing
seeds and/or excised radicles on MS medium with 2 mg/
1 2,4-D. Callus induction occurred within two weeks.
Each anther culture also developed light yellow fri-
able callus after three weeks of incubation on N_6
medium containing 2 mg/1 2,4-D in almost all genotypes. In general, callus formation from anthers, seeds and radicles was uniform within a particular genotype, but remarkable genotypical differences were
observed in callus induction ability and callus growth rate. The detailed data will be published elsewhere.

Effect of streptomycin on callus proliferation. The growth
of rice callus cultures was inhibited by streptomycin sulphate concentrations greater than 250 µg/ml

(Fig. 1), as we found before for alloplasmic lines of wheat (Kinoshita & Mikami, 1984). The medium containing 1000 or 1500 μg/ml streptomycin almost completely prevented the proliferation of the inoculated calli, which at these concentrations turned brown or black in the majority of the genotypes used. On the contrary, callus cultures at 250 or 500 μg/ml streptomycin enabled us to detect a wide range of genotypical differences in resistance to this drug. In the case of cultures in 500 μg/ml streptomycin, the resistance index varied from 0.752 to 0.058, depending on the genotype (Table 1). Additionally, variation in resistance to this drug seemed to be somewhat greater in *indica* than in *japonica* genotypes.

Similar results were also obtained when testing radicle derived callus, and a positive correlation (r = 0.450, P < 0.01) was observed between resistance indices of seed and radicle derived cultures. On the other hand, the streptomycin concentration needed for significant inhibition of callus growth in anther cultures was considerably lower than those for the other cultures (Table 1). This may be due to the higher proportions of haploid cells involved in the anther callus tissues (Oono, 1975). Though definite proof cannot yet be given, resistance indices of anther cultures were not correlated with those of seed and radicle derived cultures in our case.

Selection of streptomycin resistant callus clones. Selection for streptomycin resistance was initiated with callus from seed and/or immature embryo cultures of two rice genotypes, the Japanese cultivar Iwakogane and a Japanese genetic tester H-21. The two genotypes were relatively sensitive and produced vigorous callus on the medium without streptomycin.

Callus cultures were exposed to various levels

of streptomycin (0 to 1000 μg/ml) in the initial and subsequent selection cycles. For each culture cycle the fastest growing sectors of calli were selected for subculture. Some of these sectors grew into small healthy callus clusters after two or three weeks; however, by the end of cycle 5, half of the cultures could not be maintained longer because of slow growth and deterioration of the callus.

By visual estimation, a total of 11 comparatively vigorous callus clones were isolated after 7 or 12 subculture passages. These potentially resistant clones were then transferred to 0 and 500 μg/ml streptomycin media to calculate the resistance index. As shown in Table 2, several callus clones (e.g. Iwakogane No. III, No. IV and No. XIV) exhibited a significant increase of index as compared to the first cycle of selection, while there was only a slight increase in unselected cultures (Iwakogane No. 1 and H-21 No. 1). Thus, streptomycin resistance appears to be increased during the selection process. After 5 or 6 selection cycles, 79 plantlets were regenerated from seed callus cultures, Iwakogane Nos. XIII and XIV. In this case, however, all regenerates were albino and failed to grow to maturity.

DISCUSSION

Our results indicated that streptomycin is effective as a growth inhibitor for rice cultures, as previously mentioned in *Nicotiana* (Maliga et al., 1973; Umiel et al., 1978), maize (Umbeck & Gengenbach, 1983) and wheat (Kinoshita & Mikami, 1984). The degree of inhibition in callus growth depended upon the genotype, drug concentration and tissue source of callus. Furthermore, the inhibitory effect made it possible to select streptomycin resistant cell lines on the medium containing streptomycin.

Regeneration of mature plants from resistant callus cultures would provide information on whether resistance is attributed to a stable genetic change, or is a result of adaptation response, such as induction of an enzyme system for inactivating the drug or repression of the uptake mechanism. In a preliminary study with seed callus cultures, however, auxotrophic mature plants have not yet been obtained from the potentially resistant clones because of the inability to form green shoots. Alternatively, green plant regeneration from immature embryo callus clones is currently under investigation.

It should be also pointed out that the present selection system was different from that employed by Maliga et al. (1973). In the latter case selection was based upon the ability to form green callus and shoots in the presence of streptomycin. As described by Dix et al. (1977), the differences in selection system may influence the types of variants which were obtained.

Streptomycin inhibits protein synthesis on the prokaryotic type ribosomes found in plant chloroplast (Maliga et al., 1980). *In vitro* selection for resistance to this antibiotic, therefore, may result in the isolation of mutants with altered organellar ribosomal proteins. For instance, a streptomycin resistant mutant SR1 of *N. tabacum* has been shown to inherit the resistance as a cytoplasmic trait (Maliga et al., 1973, 1975) and to be characterized by the alteration of chloroplast ribosomal proteins (Yurina et al., 1978). Further investigations will be required to determine the exact nature of the processes which confer the resistance.

II.   Somaclonal selection of salt tolerance as an
      application of mutagenesis

INTRODUCTION

Screening techniques for selecting salt tolerant lines is an important means to generate genetic
variability. Tissue cultures have been carried out
in an attempt to increase the salt tolerance but the
results were generally unsuccessful (Bright 1985).
At a whole plant level, it is widely known that mutagenic treatments are desirable to increase the frequency of genotypic changes. Therefore, mutagenesis
in tissue culture and mutagenic consequences of the
tissue culture cycle must be investigated to develop
an efficient method of *in vitro* selection of salt
tolerance in cereals.

In this experiment, the authors first explored
the culture procedures for salt tolerance. Further
experiments are carried out on mutagenesis by using
the tissue culture method.

MATERIALS AND METHODS

Three cultivars of *japonica* rice, Shiokari,
Norin 8 and Taichung 65 were used in the experiments.
In the first experiment (Fig. 2), seed callus possibly derived from scutellum was induced by using
Chu's medium supplemented with sucrose at 50 g/l and
2.4-D at 4 mg/l. For the selection medium, NaCl
(0, 1.0 and 1.5%) was added. Tolerance of callus
was evaluated by the viability rate of the callus
after 100 days from the plating of the selection
medium. In the second experiment (Fig. 3), seed
calli were induced by the same agar medium used in
Experiment 1, and then these were transferred to

the suspension culture to propagate the calli. The
solution is B5AA which were used for rice protoplast
culture by Toriyama and Hinata (1985). The proli-
ferating calli were sieved into pieces from 0.35 mm
to 1.0 mm and transferred to the selection medium
with Na Cl. After the first selection, the surviving
calli at 1.0% were shifted to the suspension culture
for the propagation. The calli were then used to
estimate the viability rate on the selection medium.

RESULTS AND DISCUSSION

Experiment 1: The outcome of callus tolerance
is shown in Table 3a. Only the cultivar Shiokari was
used for the first experiment. The survival rates of
calli decreased prominently in higher concentrations
of NaCl.

Experiment 2: In the primary selection, the
maximum concentration of NaCl was 1.0% throughout the
three cultivars. Then, suspension culture without
NaCl was used for the proliferation of calli which
were transferred from the medium of 1.0% NaCl.

As shown in Table 3b, the calli from the three
cultivars used in the second selection survived even
at the 1.5%. From the comparison with the results
of Exp. 1, it is evident that the calli via the primary selection increased the salt tolerance. In addition, the genotypic difference among the three cultivars was also recognized. Thus, the repeated selections are effective to strengthen the tolerance.
However, it is uncertain whether the increase of the
tolerance is due to the genotypic change or the metabolic adaptation of cultured cells. It is our aim
in further experiments to select normal, fully fertile
salt tolerant regenerants with the aid of *in vitro*
mutagenesis.

REFERENCES

[1] BRIGHT, S.W.J., Selection in vitro. In Cereal Tissue and Cell Culture. In Bright, S.W.J., Jones, M.G.K., Eds., Martinus Nijhoff/ Dr. W. Junk Publishers, Dordrecht (1985) 231.

[2] CHU, C. et al., Establishment of an efficient medium for anther culture of rice through comparative experiments on the nitrogen sources. Sci. Sin. 18 (1975) 659.

[3] DIX, P.J., JOÒ, F. MALIGA, P., A cell line of *Nicotiana sylvestris* resistance to kanamycin and streptomycin. Mol. Gen. Genet. 157 (1977) 285.

[4] GALUN, E., AVIV, D., Cytoplasmic hybridization: Genetic and breeding application. In Handbook of plant cell culture. Vol. 1. EVANS, D.A., SHARP, W.R., AMMIRATO, P.V., YAMADA, Y., Eds. McMillan Pub. Co. New York, (1983) 358.

[5] KINOSHITA, T., MIKAMI, T., Alloplasmic effects on callus proliferation and streptomycin resistance in common wheat. Seiken Ziho, 32 (1984) 31.

[6] KIRK, J.T. O., TILNEY-BASSET, R.A. The plastid. Elsevier/North Holland Biomedical Press, Amsterdam, (1978) 676.

[7] MALIGA, P., Streptomycin resistance is inherited as a recessive trait in a *Nicotiana sylvestris* line. Theor. Appl. Genet. 60 (1981) 1.

[8] MALIGA, P., SZ-BREZNOVITS, A., MARTON, L., Streptomycin resistant plants from callus culture of haploid tobacco. Nature New Biol. 224 (1973) 29.

[9] MALIGA, P., SZ-BREZNOVITS, A., MARTON, L., Non-Mendelian streptomycin resistant tobacco mutant with altered chloroplasts and mitochondria, Nature 255 (1975) 401.

[10] MALIGA, P. et al., Antibiotic resistance in *Nicotiana*. In Plant cell cultures: results and perspectives. SALA, F., PARISI, B., CELLA, R., CIFFERI, O. Eds., Elsevier, Amsterdam, (1980) 161.

[11] MURASHIGE, T., SKOOG, R., A revised medium for rapid growth and bioassays with tobacco tissue cultures. Physiol. Plant, 15 (1962) 473.

[12] ONO, K., Production of a haploid plant of rice (*Oryza sativa*) by anther culture and their use for breeding. Bull. Natl. Inst. Agr. Sci. Ser. D 26 (1975) 139.

[13] TORIYAMA, K., HINATA, K., Cell suspension and protoplast culture in rice. Plant Sci. 41 (1985) 179.

[14] UMBECK, P.E., GENGENBACH, B.G., Streptomycin and other inhibitors as selection agents in corn tissue cultures. Crop Sci. 23 (1983) 717.

[15] UMIEL, N., Streptomycin resistance in tobacco: III. A test on germinating seedlings indicates cytoplasmic inheritance in the St-R 701 mutant. Z. Pflanzenphysiol. 92 (1979) 295.

[16] UMIEL, N., BRAND, N.W., GOLDNER, R., Strepto-mycin resistance in tobacco, I. Variation among calliclones in the phenotypic expression of resistance. Z. Pflanzenphysiol. 88 (1978) 311.

[17] YURINA, N.P., ODINSTOVA, M.S., MALIGA, P., An altered chloroplast ribosomal protein in a streptomycin resistant tobacco mutant. Theor. Appl. Genet. 52 (1978) 125.

Table 1. Streptomycin resistance in rice tissue cultures.

Culture	Strain	Resistance index (500 µg/ml streptomycin)								No. of strains
		0--0.1	--0.2	--0.3	--0.4	--0.5	--0.6	--0.7	--0.8	
Seed	japonica	1	11	23	7		1			43
	indica	1	2	4	9	2	1		2	21
	japonica x indica		2	3	2					7
	Total	2	15	30	18	2	2		2	71
Radicle	japonica	3	18	27	9	1				58
	indica	1		4	6	3	4	4		22
	japonica x indica	2	2	1	4					9
	Total	6	20	32	19	4	4	4		89
Anther	japonica	16	19	1						36
	indica	7	4	2	1					14
	japonica x indica	1	1	2						4
	Total	24	24	5	1					54

Table 2. Selection of streptomycin resistant callus clones in rice.

1) Callus from immature embryo culture

Strain	Clone	Concentration of streptomycin (μg/ml)			Resistance index[c] after final passage
		T-1[b]	T-2	T-3 to T-7	
Iwakogane (0.136)[a]	I	0	0	0	0.338
	II	500	0	0	0.583
	III	1000	250	250	0.765
	IV	250	0	250	0.810
H-21 (0.058)[a]	I	0	0	0	0.246
	II	500	500	500	0.249
	III	1000	0	0	0.274
	IV	500	0	500	0.632
	V	1000	250	250	0.633
	VI	250	250	250	0.634

2) Callus from seed culture

Strain	Clone	Concentration of streptomycin (μg/ml)							Resistance index[c] after final passage
		T-1	T-2	T-3	T-4	T-5	T-6	T-7 to T-12	
Iwakogane (0.130)[a]	XII	250	250	500	250	0	250	250	0.252
	XIII	250	500	250	250	500	1000	0	0.624
	XIV	250	1000	1000	250	250	250	250	0.730

[a] Resistance index of primary callus.　[b] Number of subculture transfer.

[c] Resistance index = $\dfrac{\text{Increase in mean fresh weight at 500 μg/ml streptomycin}}{\text{Increase in mean fresh weight at 0 μg/ml streptomycin}}$.

Table 3.   Selection of salt (NaCl) tolerant callus

a)   Experiment 1.

Cultivar	Treatment	Total Calli	Survived	Died	Survival(%)
Shiokari	0%	500	478	22	95.6
	0.5	500	310	190	62.0
	1.0	500	104	396	20.8

b)   Experiment 2.

Cultivar	Treatment	Total Calli	Survived	Died	Survival(%)
Shiokari	0	250	249	1	99.6
	1.0	250	157	93	62.8
	1.5	250	9	241	3.6
Norin 8	0	250	243	7	97.2
	1.0	250	60	190	24.0
	1.5	250	6	244	2.4
Taichung 65	0	250	250	0	100.0
	1.0	250	198	52	79.2
	1.5	250	48	202	19.2

94

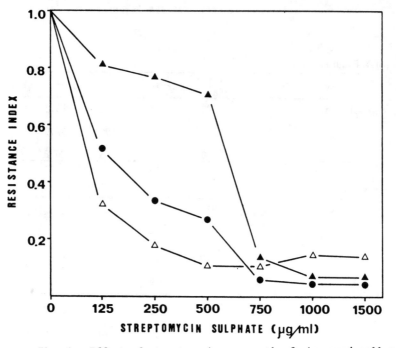

Fig. 1. Effect of streptomycin on growth of rice seed callus.
▲:I-54(*indica*); ●:A-58(*japonica*); △:E-25 (*indica*).

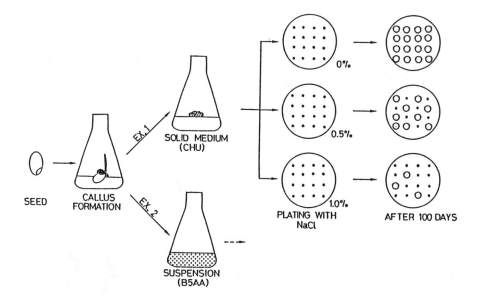

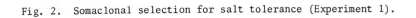

Fig. 2.   Somaclonal selection for salt tolerance (Experiment 1).

96

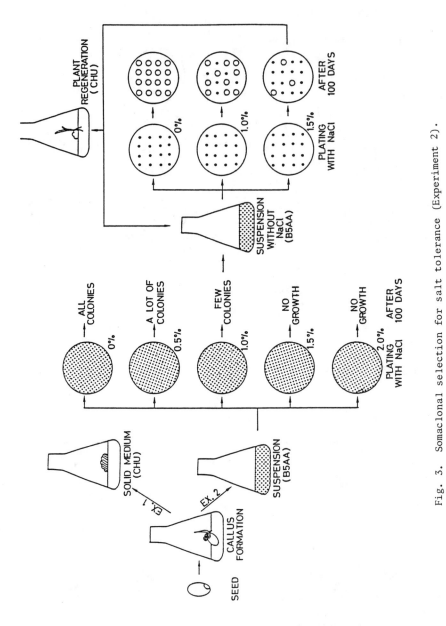

Fig. 3. Somaclonal selection for salt tolerance (Experiment 2).

EXPLOITATION AND ANALYSIS OF HETEROSIS IN WHEAT WITH INDUCED MUTATIONS

C.F. KONZAK

Department of Agronomy and Soils
and Program in Genetics and Cell Biology
Washington State University
Pullman, Washington, USA

*Scientific Paper No. 7724. College of Agriculture and Home Economics
Research Center, Washington State University, Pullman, Project Nos. 1568,
1570.

Associated with the IAEA under Research Agreement No. 4468/CF.

M. Maluszynski (ed.), Current Options for Cereal Improvement, 97–113.
© 1989 by Kluwer Academic Publishers.

98

Abstract

EXPLOITATION AND ANALYSIS OF HETEROSIS IN WHEAT WITH INDUCED
MUTATIONS

A diallel analysis of hexaploid wheat F_1 hybrids was conducted using three semidwarf mutants and three near-isogenic
semidwarf lines, all having the genetic background of the Burt
cultivar. Expressions of heterosis exceeding the high parent for
plant height and grain yield were demonstrated. However, only in
a few cases did even the F_1 height exceed that of the high
parent, and in no case did yields exceed that of the initial
cultivar, Burt. Thus, the results could be explained on the
basis of additive or dominant gene actions.

In contrast, the induced mutants were found to produce
heterosis for grain yield in the F_1 generation of tester crosses.
Grain yields of some mutant x tester F_1 combinations significantly exceeded that of the best parent, the initial cultivar
Burt, as well as all F_1 tester combinations made with
backcross-derived normal tall or semidwarf near-isogenics of
Burt. Thus, the studies showed that genetic variability carried
by the induced semidwarf mutants was responsible for increasing
the vigor of the F_1 hybrids. The results in this study may have
been limited by the small sample of mutant lines tested, since in
other crops only a part of each collection of mutants has proved
to produce heterosis in F_1 combinations. In those cases analyzed
by others, heterosis sometimes was related to the interaction of
specific genes or alleles. This specific combining ability type
of heterosis is often neglected by breeders in favor of general
combining ability, even though the two types may be of independent origin and action. The results suggest that induced
mutations can be effective for increasing the practical levels of
heterosis exploitatable via hybrid crops, and even more
important, can be exceptional tools for investigating the physiological genetic basis for heterosis phenomena.

In practice, the exploitation of heterosis by plant breeding
may be more routine than indicated via the development of hybrid
cultivars. Certainly, the fact of heterosis is readily demonstrated for those crops in which hybrid cultivars are now common.
But, perhaps we have not recognized our exploitation of the
phenomenon in the development of homozygous self-pollinated crop
cultivars. This may be true even if F_1 heterosis should become
more easily exploitable in these species.

The physiological genetic basis for hybrid vigor, the
"luxuriance of growth" which defines the heterosis phenomenon
according to the original concept of Shull [1], is still poorly
understood, even though considerable evidence and support has

been accumulated for the "dominance gene" hypothesis of East and Jones [2].

This paper will present evidence of induced mutant heterosis in wheat obtained some years ago at Washington State University, then consider this evidence in relation to work reported by others and to progress in plant breeding.

MATERIALS AND METHODS

Materials: A diallel cross was made with three reduced height mutants and three near-isogenic lines classified as very short (VS), semi-short (SS), or tall (T) from the cross CI13253// 6*Burt. Also, five reduced height mutants and five near-isogenic reduced height lines from each of three groups were crossed with two common tester stocks (Table I). The genetic background and interrelationships of the reduced height lines and testers is as follows: The induced mutant wheat lines used in the research described here were derived from gamma radiation treatments applied to seeds of the hard white winter wheat cultivar, Burt,· CI12696 (Table I). The mutants were selected in the M_2 generation and increased from M_3 homozygous lines without further selection. The set of mutant lines used in the study represented the range of plant height levels observed among a group of more than 100 mutants, of which only about half have been maintained. Several are available in the USDA Collection.

The reduced height gene from the shortest mutant, Burt 937 (CI15076), was later found to act as a complete recessive and was designated rht4 [3]. The Burt 818 mutant also carries a single recessive gene controlling the difference in height expression from Burt and the gene is independent of rht4 (Konzak and Hu, unpublished). The genetic basis of reduced plant height in the other mutant lines has not been determined.

Reduced height lines near-isogenic for one or both Rht1 and Rht2 were included to represent semidwarfing sources of spontaneous mutant origin (Table I). The near isogenics in Burt background were then in the process of development by Dr. R. E. Allan, USDA-ARS Wheat Geneticist, at Pullman. The Rht gene(s) carried by the near-isogenic stocks were not then known with certainty, especially those carrying Rht1 or Rht2. All of the mutants and some of the near-isogenic stocks carry rht6 in addition (see C. F. Konzak, 1987 [4] for listing of Rht genes). However, the very short near-isogenic WA5490 most certainly carries Rht1 + Rht2 and perhaps rht6. The rht6 gene is responsible for the small (10-15 cm) height difference between Burt and Itana, as was determined from F_2 and F_3 progeny analyses (Konzak and Hu, unpublished).

Of the tester lines, Sel 101 (CI13438), was derived from the cross CI12353/CI12697. CI12353 is the original Norin 10/Brevor, Sel 14 which carries Rht1, Rht2. CI12697 was a sister selection to the mutant and near-isogenic background source Burt (CI12696).

Itana (CI12933) was selected from the cross Blackhull/Rex//
Cheyenne, hence could only be very distantly related to Burt.
Itana and CI13438 were selected as testers based on previous work
of R. E. Allan (unpublished) showing their high combining
ability. CI13438 is now known to carry Rht1, but probably not
rht6. Itana carries no semidwarfing genes into the combinations.

Methods: The F_1 seeds and check varieties were sown in three
experiments on land previously fallowed after three years of
alfalfa plus grass cut for hay. The seeds for the "diallel" were
sown by hand 1.5 cm apart in rows 0.92 M long x 30 cm wide in
accordance with a completely randomized block design with 14
replications. Those for the tester crosses were sown 10 cm apart
in similar rows, but there were eight replications.

Data were collected from direct measurements in the field
and in the laboratory on pulled plants sampled randomly from each
replicate and treatment. Data were recorded for plants/plot,
total tillers with fertile spikes per plot, adult plant height,
number of fertile spikelets on main tillers (2 pl/plot), sheaf
weight, grain yield, plant height components, plant weight,
tillers/plant, yield/plant, kernels/spike, and 100 kernel weight.

Data were subjected to analyses of variance, and to
combining ability analyses based on Griffings fixed Model I.
Heterosis values were calculated for all variables, using the
mean of the F_1s and parents in each cross. F_1 heterosis was
calculated also with respect to the high parent mean of the cross
combination, "heterobeltiosis", according to Fonseca and
Patterson [5, 6].

RESULTS

The research described here was conducted at Pullman,
Washington, during 1966-69 as a Washington State University
graduate research project [7]. Part of the results were
published earlier [8]. The objective of the research was to test
the possibilities for heterosis effects using intercrosses among
reduced and normal height mutant and near-isogenic lines with a
generally common genetic background, and to compare the results
obtained with crosses to two tester lines having divergent
genetic relationships to the near-isogenic and mutant stocks.

The data from this study are too voluminous to consider in
detail here. Only excerpts pertinent to the objectives of this
program will be presented. Further information is available in
Ramirez [7] and Ramirez et al. [8].

Diallel crosses - Data from the crosses involving mutants
and near-isogenic lines showed that either compensation or
dominant epistatic effects on plant height occurred among the
various combinations. In no case was the F_1 plant height greater
than that of the initial cultivar/recurrent parent, Burt (data
not shown).

In all crosses, the Rht1 Rht2 genotype caused a partially additive reduction of F_1 plant height (data not shown). Height expression in the mutant/mutant combinations was not greatly different from that expected for genetic compensation in a digenic F_1 recessive cross combination, although the actual means for one combination suggested partial dominance of the mutant Rht gene on plant height.

Tillers/plant and spikelets/spike did not differ for any mutant combination among parents. Grain yield/plant varied, with the highest values obtained for the tall mutant and initial cultivar (Burt) while lowest values were obtained for the Rht1 Rht2 isogenic. None of the F_1 combinations among "isogenic" lines produced grain yields above that of Burt, although several did not differ significantly from it. The 100 kernel weight most closely paralleled the grain yield values, indicating that kernel weight was the main differentiating yield component in the experiment. Kernel weight was inversely related to plant height among both the near-isogenic lines and the mutants, while the grain weight of even the shortest mutant was slightly higher than that of the tall near-isogenic parent (data not shown).

Generally 100 kernel weights of the F_1 were intermediate between the parents, except for the tall near-isogenic/medium height mutant cross, which showed significant "high parent" heterosis, yet not full complementation for kernel weight. Interestingly, significant heterosis occurred for increased number of kernels/spike in the short isogenic/short mutant cross, but the 100 kernel weight (2.6 vs 2.9 g) for the F_1 was nearer that of the higher kernel weight mutant parent, much above the near-isogenic parent (0.9 g). The net effect, however, was insufficient to reach the yield of the Burt parent.

Variation in most other parameters measured seemed explainable on the basis of simple dominance or additive effects, due to genetic complementation, since the CI12353/6*Burt near-isogenics and the mutants both approximate Burt in their genetic composition.

Tester crosses: For the mutant and near-isogenic line/tester crosses, the influence of the mutant semidwarfing genes on F_1 plant height would not have been predictable without at least some knowledge of the semidwarfing gene inheritance, and these were the first crosses made to obtain that information. As indicated earlier, we later found that CI15076 (Burt M937) and WA4564 (Burt M818) each carry different single recessive mutant genes in addition to the common rht6 factor from the Burt background itself. In each of the tester cross combinations, the mutants affected plant height much as would the initial variety Burt, indicating that in general, genetic compensation occurred. Thus, the semidwarfing genes of these lines apparently had no influence on the tester cross F_1 height (Table II). Surprisingly, the effect of Rht1 in CI13438 is subdued, suggesting that some heterosis effect on plant height may occur in its tester crosses to these mutant lines. In contrast, crosses of WA5455 (Burt M1723) appeared to show some measure of additive or dominant action on plant height even in the combination with Itana. In the cross of WA5455 with CI13438, the F_1 was only slightly taller than the tester parent. These results support indications from the diallel crosses that WA5455 carries a partially dominant reduced height gene. Mutants WA5458 (Burt-1088) and WA5489 (Burt-1789) acted much as expected for Burt. In earlier studies, Burt-M1088 showed a slight (10 cm) height reduction from Burt. Burt-M1789 orginally was selected for increased height over Burt, but environment affects its expression.

The test cross F_1 grain yield data were distinctly more revealing of heterosis effects than were the plant height data (Table III). In the tester crosses with Itana, all five mutants produced F_1 that were significantly higher yielding than those involving all five Norin 10 derived near-isogenic lines. The very lowest yielding F_1 were made with the very short near-isogenics which must carry Rht1 + Rht2. Notably these F_1 had the lowest kernel weights of the series of tester cross combinations. In the mutant/Itana tester crosses, Burt-M1723 caused the lowest level of heterosis. Otherwise, all except the tallest mutant had similar effects. The highest yielding F_1 with the Itana tester was made with the tallest mutant.

Likewise, for the tester crosses between the Norin 10-derived isogenics and CI13438 (Sel 101), there was no instance of positive heterosis expression for yield. In fact, the CI13438 crosses with lines apparently carrying both Rht1 and Rht2 had yields significantly lower than CI13438.

A notably different pattern was observed for the mutant crosses to CI13438. No mutant x CI13438 F_1 combination yielded significantly less than the tester parent, while the semi-tall and tall mutant lines both invoked heterotic performance significantly exceeding the values for CI13438 or any other F_1 cross combination. Only among the mutant x tester crosses were there improvements in harvest index compared to the parental values (Table III). The relatively high harvest index of CI13438 was notably improved in its F_1 with WA4564 (Burt M818).

These results demonstrate that genetic changes carried by the mutants are responsible for statistically significant expressions of F_1 heterosis for grain yield. The observed grain yield values exceeded not only those of the initial cultivar in which the mutants were induced, but also those of the testers used in the crosses analyzed.

DISCUSSION

Most of the available plant data on heterosis introduced in crosses by induced mutants has come from studies with diploid species. The most extensive data on the subject comes from research on barley. However, it is perhaps important to note that possibly the first report of single gene heterosis was in maize by D. F. Jones [9] for a deleterious spontaneous mutation. Reports of a similar result for an induced chlorophyll lethal mutation in barley by Ake Gustafsson [10, 11] brought into greater focus this seeming relation of deleterious mutations to heterosis. In fact, a recent study in Argentina by Salerno et al. [12] indicates that the yielding capacity of a synthetic maize population is best maintained by retaining rather than removing the spontaneous chlorophyll-deficient mutants which inevitably occur. While the association between mutant defects and heterosis may have some as yet unknown significance, Doll [13] has convincingly established that heterosis is not commonly associated with heterozygosity for chlorophyll deficiency mutants in barley.

Thus, as is evident also in other work, heterotic effects are characteristic of specific mutant genotypes and cross combinations. Gustafsson et al. [14, 15] have demonstrated heterotic effects related to allelic interactions at a single erectoides mutant locus in barley. In this case the mutant genes themselves have breeding value separate from the heterotic interactions observed.

The very extensive investigations of Maluszinski [16] and Maluszinski and colleagues [17] in Poland have shown that an abundance of mutants induced in barley by nitrosourea (MNUA) express F_1 heterosis for yield at levels exceeding the initial cultivar or best parent either from mutant/cultivar or mutant/mutant crosses. In these studies, there seems no clear pattern or association of heterosis effects with specific mutant traits, allowing any measure of predictability. However, because most of the mutants studied were selected first for their reduced height, these results support the view that heterosis can be associated with mutant types of high breeding value. Interestingly, while high yielding mutants produced heterotic F_1, high combining ability also was found for mutants ranging widely in their "viability" and performance. The trait most commonly associated with high yield was increased tillering, often without reduced kernel weight. But, in some F_1 combinations, tillering was unaffected while grains/spike and kernel weight were increased. In the studies by Maluszinski [16, 17], F_1 heterosis was observed for mutant/mutant or mutant/initial variety combinations within the same genetic background, as well as for mutant/mutant or mutant/cultivar combinations differing in genetic background. Some mutants contributed to heterosis in several different F_1 combinations, whereas the effectiveness of others was more limited, indicating that the mutants possessed different levels of specific combining ability. Even negative heterosis for grain yield was observed for combinations in which the two parents showed high positive heterosis for grain yield in other crosses.

Studies by Gottschalk [18, 19], in peas, suggest that heterosis effects for yield may often be associated with fasciation mutants, although certain mutants appear to be more effective than others. It is not clear from his studies whether any other mutants have similar combining ability, or if further investigations are needed.

Luxuriant growth leading to high forage yield was found due to heterozygosity for induced mutants in sweet clover (Melilotus alba) crosses [20]. Micke [20] also found that the vigor of the F_1 generation of crosses even among low yielding sweet clover mutants could be increased above that of the initial line. Micke proposed a breeding scheme that would exploit heterosis while allowing the development of coumarin-free sweet clover cultivars.

Heterosis for increased vigor and branching also has been observed in Petunia axillaris mutants by Gorny [21].

The results for hexaploid wheat mutants described earlier are in many ways comparable to those reported for diploid species. The wheat work, however, has involved only a very small sample of the mutants actually induced, so it may even be surprising that F_1 heterosis for yield or other traits was detected at all in the crosses. Nevertheless, the results indicate that significant, useful F_1 heterosis effects can be expected from hexaploid wheat mutants. Moreover, the broad range heterotic gene interactions may be more exploitable in polyploids than in diploids. MacKey [22] has often pointed out that the genetic redundancy present in polyploid wheats provides a basis for mutation of alleles to new functions, or deletion, such that interallelic heterosis such as described by Gustafsson [10, 11] and Gustafsson et al. [14, 15] could occur between homeologous duplicate genes and fixed in homozygous lines. The common occurrence of various genetic lethals (D_1-D_4 Ch 1, Ch 2, Ne_1, Ne_2) in wheat populations over the world (See Hermsen [23]; Worland and Law [24]; Zeven [25]) suggests that these lethals/semilethals/"defectives" may serve a function rather than be a "genetic load" in the terms of Muller [26]. The fact that wheats in some areas do not carry one or other of the "defective" loci is not important. If the "defective" loci did not have at least neutral value they would likely be eliminated by natural selection.

It has previously been suggested that there is a fundamental basis for reduction in redundant genes in polyploids [27]. In polyploid wheats, the two (tetraploid wheats) or three (hexaploid wheat) genomes carry homoeologous redundant loci for many genes. As all genes carry out their activities via cytoplasmic organelles, it is important to note that the two or three sets of homoeologous genes must carry out these functions via cytoplasm of a singular origin [27]. A prime focal point for this argument is ribulose diphosphate carboxylase (RudpCase) or (Rubisco) for which the large subunits are controlled by the chloroplast genes, while nuclear genes determine the small subunits. The structure and thus the efficiency of the Rubisco in a wheat genotype may be differentially affected by alterations in the subunits under nuclear control. If the two or three homeologous pairs of genes controlling the subunits are identical, the redundancy may cause metabolic inefficiency unless as yet unknown feed-back control systems are present. However, if the function of any one or more of these homeologous nuclear genes is modified by mutation, but still able to compete for a site in the Rubisco molecules, the result could be reduced/increased efficiency of energy storage by the genotype even in the presence of a feed-back control system. In this example, mutations might also change the function of such a gene, or delete it, as a basis for increasing Rubisco efficiency.

Certainly, the possibilities for mutational improvement of wheat are considerable, and in fact, the "diploidization" which is believed to occur as the result of the plant breeding progress [22] -- "experimental evolution" -- to be sure, is the result of selecting gene alternatives for either redundant or inefficiently functioning loci. Thus, it is not surprising that a portion of the mutant variations induced can improve the function of a hybrid genotype. Whether such changes can be genetically fixed also in diploids may depend on the degree of redundancy present for genes controlling the processes affecting plant vigor, especially yield, via yield components.

DIRECTIONS FOR THE FUTURE

Although in specific instances it has been shown that heterosis effects are due to the interactive expression of alleles at a single locus, there are now numerous examples of mutant induced hybrid vigor for which the genetic basis is not yet resolved. Important in this respect is the fact that these mutants can be exploited as tools with unusual precision for investigating the physiological genetic basis for heterosis. To improve the isogenicity of the background genotype for use in these investigations, the mutants can and should be backcrossed to the initial line to remove unwanted secondary alterations. The genetic variability among backcross progeny should then be sampled via test crosses to determine if any visible and heterotic mutant gene effects can be separated.

Progeny lines selected from the initial variety/mutant and reciprocal backcrosses (as may be necessary to test for cytoplasmic changes) once proved to carry the heterotic loci can provide ideal material for investigations into the physiological genetic basis for the heterosis phenomenon. Studies of Gustafsson et al. [14, 15] already have offered new insight into the problem, but should be followed in further depth. Recent work by Dollinger [28] in maize also needs to be followed up using lines derived from backcrosses to the initial inbred in which the mutants were induced. With wheat, similar studies are essential, especially since the polyploid wheat genomes more readily absorb genetic alterations, many of which may be unrelated to the trait of interest. This was certainly the case with a new chocolate brown chaff color mutant induced in durum wheat recently, as it was found possible in F_3 and F_4 progenies to separate the single recessive chaff color trait from the partial sterility-shrunken seed condition characteristic of the original isolate. A loose linkage with the chaff color gene or large genetic deficiency may have been involved, and the color expression is not as distinct in progenies with normal seeds. Using a new genetic method of analysis we found that the new chaff color gene is located on chromosome 7B (Joppa and Konzak, unpublished). The mutation apparently involves a modification of anthocyanin pigmentation which is sensitive to temperature and the C:N ratio, thus it is not surprising that expression of the trait was stronger in the original line.

As has been demonstrated by Gottschalk [19] in peas, genetic analyses can be effective for identifying mutant loci responsible for heterotic effects. This is more feasible because the background of mutants is much the same as the initial stock in which they were induced. In wheat, the availability of numerous cytogenetic tester stocks (ditelosomics, etc.) offers the possibility to test the hypothesis of fixed heterosis via alteration of intergenomic genetic redundancy.

As most of the field studies on mutant heterosis have been done under conditions favoring low interplant competition, emphasis now needs to be placed on studies utilizing the recently developed chemical hybridizing agents, or CHAs, to produce sufficient hybrid seed for standard field trials conducted under a range of environmental conditions. It might be suspected that increases in some plant parts due to heterosis may be less or more important under more competitive situations. There may be environmental limits, for example, to increases in tillering, or to increases in reproductive organs, such that the heterosis observed under spaced planting will not occur under more usual field plant densities. On the other hand, if commercial F_1 hybrids of small grains and grain legumes are to become a reality, the expected higher cost of the hybrid seed may be best compensated for by greater efficiency in the F_1 hybrids to exploit space available, yet achieve higher economic yields. Thus, it will be important to study effects of seeding rates and seed distribution (row spacing) on the productivity of the hybrids grown in standard nursery trial plots. Fortunately, most semidwarfing mutants induced in wheat are semidominant, while one, Rht12 is dominant, and consequently are effective for controlling height expression in the F_1 generation, a definite asset for hybrid cultivar development.

Further into the future, one can envision incorporating by genetic engineering techniques and thereby fixing heterotic mutant genes into responsive genotypes, whether they be diploid or polyploid. And, perhaps even more far-fetched is the idea that once we have greater knowledge of the physiological genetic basis of heterosis, genetic engineering techniques can be used to construct selected genetic elements in vitro prior to inserting them in the genome where they are to function.

The techniques discussed here relate to the practical exploitation and analysis of heterosis using induced mutation techniques. It must be recognized, however, that the approaches discussed are not in themselves sufficient basis for continued increases in plant productivity. No one method, even modern genetic engineering, can be expected to bring about all the genetic modifications even now envisioned as necessary to achieve the production potential of each crop. However, induced mutations surely will play an important role in efforts toward that goal.

PLANS FOR FUTURE RESEARCH
C. F. Konzak

I. Heterosis

a. Investigations using CHAs will continue in efforts to determine effective treatments. This work will be done using hexaploid wheats, durums, and possibly oats.

b. Lines having potential for use in hybrids will be selected and increased for use in 1988.

c. Additional mutant stocks will be selected and increased for possible use in 1988.

d. Development of near isogenic stocks carrying Rht12 or Rht13 will be continued, and phenotypically improved lines carrying Rht12 will be selected from winter and spring habit cross progenies.

e. Limited tester crosses will be made for preliminary investigations of heterosis prior to use of CHAs.

f. Genetic analyses of new semidwarf mutants will be continued (private funding), crosses will be made to transfer Rht12 to tetraploid wheats.

g. Backcross-derived lines of durum will be increased for testing the association between the Rht16 or Rht18 mutant genes and F_1 heterosis (private funding).

II. Haploids and tissue culture

a. Studies to produce haploids from anther culture will be initiated as may be feasible (with a new trainee).

b. Possibilities for application of mutagens to anthers in culture, and to tissue cultures will be investigated, at least via literature review and contacts with other laboratories.

TABLE I. NEAR ISOGENIC AND MUTANT LINES IN BURT BACKGROUND

Acc. No.	Pedigree/ Ident.	Probable Rht/gene	Mean plant height (cm)	Class[a]
WA5490	CI13253/6*Burt	Rht1,+ Rht2,+ rht6	47	VS
WA5491	CI13253/6*Burt	Rht1,+ Rht2	51	S
WA5492	CI13253/6*Burt	Rht1 + rht6?	69	SS
WA5518	CI13253/6*Burt	rht6	93	MT
WA5493	CI13253/6*Burt	--	96	T
CI15076	Burt-M937	rht4, rht6	63	VS
WA5455	Burt-M1723	1 partial rec?, rht6	79	S
WA4564	Burt-M818	1 rec, rht6	80	SS
WA5458	Burt-M1088	?, rht6	89	MT
WA5489	Burt-M1789	?, rht6	103	T
CI12696	Burt	rht6	92	T

[a] Class   VS = Very Short, S = Short, SS = Semi-short, MT = Medium Tall, T = Tall

TABLE II. PLANT HEIGHT IN TEST CROSSES

Parent ID.	Parent ht.[a]	Itana[a] Cross	DMRT[b] P<0.05	CI13438[a] Cross	DMRT[b] P<0.05
WA5490	47	86	ab	64	a
WA5491	51	90	bc	64	a
WA5492	69	100	d	78	bc
WA5518	93	110	ef	89	d
WA5493	96	114	f	93	de
CI15076	63	114	f	90	d
WA5455	79	106	e	82	c
WA4564	80	111	f	89	d
WA5458	89	112	f	90	d
WA5489	103	115	f	95	e
PARENTS					
Burt	94		c		d
Itana	112		f		
CI13438	80				c

[a] height in cm.
[b] Duncan multiple range test. Values with same letter are not significantly different.

TABLE III. F_1 GRAIN YIELD[a], HARVEST INDEX[b] FOR TEST CROSSES

Parent ID	Parent[c]	Itana[a] Cross	DMRT[d] <0.05	HI[b]	CI13438[a] Cross	DMRT[d] <0.05	HI[b]
WA5490	2.6	7.0	ij	.37	12.0	fg	.40
WA5491		9.1	ghi	.31	13.3	ef	.41
WA5492	4.2	10.0	gh	.35	15.4	de	.40
WA5518		9.7	gh	.34	15.4	de	.37
WA5493	8.9	10.1	gh	.32	17.8	bcd	.39
CI15076	9.3	15.8	bcd	.34	16.5	cde	.39
WA5455		13.6	def	.40	14.7	cd	.43
WA4564	11.2	16.4	ab	.41	18.0	bc	.45
WA5458		15.4	cde	.39	22.8	a	.41
WA5489	13.9	18.5	a	.41	22.0	a	.42
Burt	9.3[c]	8.5	gh	.32		jklmno	
Itana	10.8		gh	.35			
CI13438	17.8			.42		bcdef	

[a] gr/plant.
[b] grain yield above ground biomass yield.
[c] Parent yield values from separate study, yield of Burt for comparison was 13.1 gr yield values for Burt, Itana and CI13438 are for comparison to F_1 yields.
[d] Duncan multiple range test for yield values. Values with same letter are not significantly different from one another for comparisons within a column of values.

REFERENCES

[1] SHULL, G.H., What is "heterosis"?, Genetics 33 (1948) 439.

[2] EAST, E.M., JONES, D.F., Inbreeding and Outbreeding, Lippincott, Philadelphia (1919).

[3] KONZAK, C.F., "A review of semidwarfing gene sources and a description of some new mutants useful for breeding short-stature wheats", Induced Mutations in Cross-Breeding (Proc. Adv. Group Vienna, 1975), IAEA, Vienna (1976) 79. STI/PUB 447.

[4] KONZAK, C.F., "Genetic analysis, genetic improvement and evaluation of induced semi-dwarf mutants in wheat", Evaluation of Semi-dwarf Cereal Mutants for Cross Breeding (Proc. 4th FAO/IAEA Res. Coord. Meet., Casaccia, 1985), IAEA, Vienna (1987) (in press).

[5] FONSECA, S., PATTERSON, F.L., Hybrid vigor in a seven parent diallel cross in common wheat, Crop Sci. 8 (1968) 85.

[6] FONSECA, S., PATTERSON, F.L., Yield component heritabilities and interrelationships in winter wheat, Crop Sci. 8 (1968) 614.

[7] RAMIREZ, I.A., Heterosis of Plant Height and Other Characters in Crosses of Mutagen-induced and Backcross-bred Semidwarf Wheat Lines, Ph.D. Thesis, Washington State University, Pullman (1969) 166 pp.

[8] RAMIREZ, I.A., ALLAN, R.E., KONZAK, C.F., BECKER, W.A., "Combining ability of winter wheat", Induced Mutations in Plants (Proc. Symp. Pullman, 1969), IAEA, Vienna (1969) 445. STI/PUB/231.

[9] JONES, D.F., Heterosis resulting from degenerative changes, Genetics 30 (1945) 527.

[10] GUSTAFSSON, A., The effect of heterozygosity on viability and vigor, Hereditas (Lund, Sweden) 32 (1946) 263.

[11] GUSTAFSSON, A., The advantageous effect of deleterious mutations, Hereditas 33 (1947) 573.

[12] SALERNO, J.C., HOPP, H.E., FAVRET, E.A., "Persistence of lethal genes in maize populations due to natural selection", Induced Mutations--A Tool for Crop Plant Improvement (Proc. Int. Symp. Vienna, 1981), IAEA, Vienna (1981) 81. STI/PUB/591.

[13] DOLL, H., Yield and variability of chlorophyll mutant heterozygotes in barley, Hereditas 56 (1966) 255.

[14] GUSTAFSSON, A., DORMLING, I., EKMAN, G., Phytotron ecology of mutant genes. I. Heterosis in mutant crossing of barley, Hereditas 74 (1973) 119.

[15] GUSTAFSSON, A., DORMILING, I., EKMAN, G., Phytotron ecology of mutant genes. II. Dynamics of heterosis in an intralocus mutant hybrid of barley, Hereditas 74 (1973) 247.

[16] MALUSZYNSKI, M., The high mutagenic effectiveness of MNUA in inducing a diversity of dwarf and semidwarf forms of spring barley, Acta Societatis Botanicorum Poloniae 51 (1982) 429.

[17] MALUSZYNSKI, M., MICKE, A., SIGURBJORNSSON, B., SZAREJKO, I., FUGLEWICZ, A., "The use of mutants for cross-breeding and for hybrid barley", Barley Genetics V (Proc. Fifth Int. Barley Genet. Symp., Okayama, 1986) In press.

[18] GOTTSCHALK, W., The productivity of some mutants of the pea (Pisum sativum L.) and their hybrids. A contribution to the heterosis problem in self-fertilizing species, Euphytica 19 (1970) 91.

[19] GOTTSCHALK, W., "Monogenic heterosis", Induced Mutations in Cross-Breeding (Proc. Adv. Group Vienna, 1975), IAEA, Vienna (1976) 189. STI/PUB/447.

[20] MICKE, A., "Hybrid vigor in mutant crosses", Induced Mutations in Cross-Breeding, (Proc. Adv. Group Vienna, 1975), IAEA, Vienna (1976) 199. STI/PUB/447.

[21] GORNY, A., Chemomutanty Petunia axillaris, Ph.D. thesis, Sitesian Univ., Katowice (cf. Maluszynski 1982), 1976.

[22] MACKEY, J., "Mutations as a means for analysing polyploid systems". Induced Mutations--A Tool for Crop Plant Improvement (Proc. Int. Symp. Vienna, 1981), IAEA, Vienna (1981) 143. STI/PUB/591.

[23] HERMSEN, J.G., "Hybrid necrosis and red hybrid necrosis", Proc. 2nd Int'l. Wheat Genet. Symp. (Lund, Sweden, 1963), Hereditas Suppl. 2 (1966) 439.

[24] WORLAND, A.J., LAW, C.N., The genetics of hybrid dwarfing in wheat, Z. Pflanzenzuchtg. 85 (1980) 28.

[25] ZEVEN, A.C., Seventh supplementary list of wheat varieties classified according to their genotype for hybrid necrosis and geographical distribution of Ne genes, Euphytica 25 (1976) 255.

[26] MULLER, H.J., Our load of mutations, Am. J. Hum. Genet. 2 (1950) 111.

[27] KONZAK, C.F., Genetic control of the content, amino acid composition, and processing properties of proteins in wheat, Adv. Genet. 19 (1977) 407.

[28] DOLLINGER, E.J., Effects of visible recessive alleles on vigor characteristics in a maize hybrid, Crop Sci. 25 (1985) 819.

BARLEY MUTANTS - DIVERSITY, GENETICS AND PLANT BREEDING VALUE

U. LUNDQVIST

Svalöf AB,
S-268 00 Svalöv
Sweden

M. Maluszynski (ed.), Current Options for Cereal Improvement, 115–128.
© 1989 by Kluwer Academic Publishers.

ABSTRACT

We have established, in barley, the existence of 12 gene loci promoting the spike development of lateral florets, their size, awn development, fertility and kernel development. One of these loci, hex-v, makes the spike form well developed kernels in six rows; the remaining ones - int loci - effect an intermediate enlargement of lateral florets, the awn development, fertility and kernel development varying in characteristic manners, intermediate between two-rowed and six-rowed barleys. Two loci, hex-v and int-d, are semidominant, with a limited phenotypic expression already in the heterozygous state. No case of indisputable linkage has been established among the 9 more intensively investigated int loci; for practical purposes, these loci may be considered to approach independent inheritance. Linkage is established between int-d and hex-v. It is intriguing that the two loci with traits of dominant/ semidominant promotion of lateral floret development should be closely linked to one another.

Tendencies to co-operation among these genes have been established, leading to enhanced promotion of lateral floret development. Thus, int genes combined in double homozygous state have frequently resulted in typical six-row spikes; such double mutants combined with the hex-v gene in triple mutants may lead to extremely beautiful six-row types; and this co-operation between the semidominant hex-v gene and recessive int genes has been distinguishable even at the heterozygous state of these genes. Finally, the presence of an int-c allele beside the hex-v gene has been demonstrated in a commercial six-row variety; this int gene evidently affects the six-row phenotype favourably. There are clear indications that the competence to co-operate efficiently depends on particular kinds of gene interaction, particular int loci being more competent than others, and the competence of the individual int allele depending on the specific constellation of loci, also with regard to the hex-v locus.

INTRODUCTION

Mutation research has provided an insight into the rather complex genetics of kernel rows in barley. Normal two-row barley, carrying, on two opposite sides of the spike, median florets with two reduced, sterile and awnless lateral florets, is able to produce six-row barley in a single mutational step. These mutants, with well developed lateral florets, fully fertile and with long awns - 41 cases so far isolated in our studies - have without exception been localized to one locus, hex-v (1), located in the long arm of chromosome no. 2 (2, 3). Besides, two-row barley may also produce mutants with spike development intermediate between the two-row and the six-row states. The enlarged lateral florets of these mutants vary in characteristic ways with regard to awn development, fertility and kernel development, not only among mutants, but also depending on environmental conditions. The 103 intermedium mutants in our studies have been localized to 11 int loci.

Such a complex situation gives a key to the somewhat confused literature with conflicting publications on the genetics of kernel rows in barley. The use of clearly defined mutants evaluated against one another in allelism tests has demonstrated that there are many degrees in the development of lateral florets in cultivated barley, both among loci and among alleles, and that genes promoting lateral floret development may interact in unexpected reinforcing or disturbing manners (4, 5). Such interactions are of great interest, both theoretically and for practical purposes.

THE INTERMEDIUM AND HEXASTICHON MUTANTS

Among our 103 gene localized intermedium mutants induced in two-row cultivated barleys, there are representatives for 11 int loci. They have been induced by several kinds of mutagens: sparsely ionizing and densely ionizing radiations, organic and inorganic chemicals. Our 69 more intensively investigated intermedium mutants, located to 9 of the loci, have been induced in the three varieties Svalöf´s 'Bonus', 'Foma', and 'Kristina'.

There are clear morphological distinctions among intermedium mutants belonging to different loci, but also among allelic mutants (4, 5). The distinction between intermedium mutants and hexastichon mutants is obvious, all our 41 six-row mutants so far obtained being localized to one and the same locus, hex-v, located in the long arm of chromosome no. 2. Among the int mutants, only the int-d alleles show dominance, which is variable among the 13 more intensively studied alleles, and also for the individual d allele.

No case of indisputable linkage has been established among the 9 more intensively investigated int loci; for practical purposes these loci may be considered to approach an independent inheritance in relation to one another. Linkage is established between int-d and hex-v. It is intriguing that the two loci with traits of dominant/semidominant promotion of lateral floret development should be closely linked to one another.

The 103 gene localized intermedium mutants are distributed on the 11 loci in the following manner:

int-a:	31 cases	int-h:	6 cases
int-b:	3 cases	int-i:	1 case
int-c:	23 cases	int-k:	1 case
int-d:	21 cases	int-l:	1 case
int-e:	14 cases	int-m:	1 case
int-f:	1 case		

Even a mutant number as large as 103 will be inadequate for demonstrating whether the different int loci show any preferences to the type of mutagen applied or whether the different kinds of mutagenic treatment display similar patterns of their distribution of mutants. It is, however, perfectly clear that the different int loci have different mutabilities, irrespectively of treatment with one or another mutagen. Thus, int-a, -c, -d, and -e embrace more than 86% of the mutants, whereas among the remaining int loci only int-b and int-h show more than just sporadic cases of mutation.

It may in this connection be relevant to put attention to our investigations of mutagen specificity in the eceriferum (cer) mutants of barley (6). Available were 1580 such mutants, localized to 79 cer loci, and induced by various mutagens (X rays and neutrons, various organic chemicals, and the inorganic sodium azide). Also for the cer loci, there is a very great range of mutabilities, irrespectively of mutagenic treatment. Thus the majority of cer loci mutate only sporadically, 44 loci having mutated from 1 to 5 times and, on the other hand, 16 of the loci embracing no less than 79% of the mutants. Moreover, it is equally obvious that different cer loci quite frequently show markedly differing mutagen specific reactions (TABLE I). Whereas ethylene imine and sulfonates show similar patterns of mutagen efficiency on the present cer loci, there are strong differences between the organic chemicals and sodium azide, which in its turn differs less from the physical mutagens X rays and neutrons than do the organic chemicals. Besides, there are clear differences between the two kinds of treatment with ionizing radiation. The wealth of alleles distributed on a large number of cer loci have provided insights into the mutation process which are, most likely, of general relevance. With regard to the int loci, not only do they show different mutabilities. Differing mutagen specific reactions should also be expected among these loci.

DOUBLE MUTANT COMBINATIONS OF INT GENES

In our almost complete tests for allelism among the representatives of the 9 more intensively investigated int loci (int-a, -b, -c, -d, -e, -f, -h, -i, and -k), the double mutant F_2 segregates rank in three groups in their spike development: six-rowed, irregular, and deformed or even abnormal spikes. This division does not show clear-cut borders among the different types. There is a continuous transition between irregular and deformed types of spike, and the demarcation is probably more or less artificial. Six-row types, on the other hand, may now and then have a some-

what irregular fertility of their lateral florets; so, even this
demarcation must be regarded with some caution. It is, however,
perfectly clear that typically six-rowed spikes and irregular
spikes always belong to different progenies, as unequivocal
indications of different kinds of interactions among different
gene loci (4, 5).

The final classification of the types of spike development
has applied a scale with 9 steps, steps 1-3 denoting six-row
types with gradually decreasing regularity, and the steps 4-6
and 7-9 corresponding to irregular and deformed spikes, respectively, of increasingly extreme shape. Means of spike phenotypes
for double mutant combinations among the int loci are entered in
TABLE II. Outstandingly most efficient combiners are int-d and
int-c.

A study of the four most frequently mutated int loci (int-a,
-c, -d, and -e) reveals characteristic differences not only among
loci, but also among their alleles in their ability to co-operate
in the formation of six-rowed spikes in double mutants (TABLE
III);in fact, a bimodal distribution of spike development values
is apparent for each of the six combinations among the four int
loci. It is remarkable that, whereas int-a and int-e both co-operate successfully with int-c and int-d, they should be so poor
partners to one another. As for individual alleles, there are no
clear signs of outstanding performances: success seems to depend
on the particular constellation of loci and alleles.

PRESENCE OF AN INTERMEDIUM GENE IN A COMMERCIAL SIX-ROW VARIETY

As recently demonstrated in our investigations, the six-row
cultivar Svalöf´s 'Agneta' carries two genes for enlarged development of lateral florets, at hex-v and int-c. Locus int-c is
apparently allelic with the so-called I^h, I, i series, which also
promotes the fertility of lateral florets in two-row barley (7).
In the control of the six-row/two-row difference, interaction
between the I^h, I, i allelic series and alleles at the hex-v locus
(v, v^d, V, v^t allelic series; (8)) has long been recognized (e.g,
9, 10, 11, 12), and even a considerable enhancement of lateral
floret fertility when either I^hI^h or II combinations accompany
the heterozygous Vv genotype (11). It has been suggested that the
genotypes vv II and vv I^hI^h both seem to occur frequently among
cultivated six-row barleys (12).

Our further analyses demonstrated, among alleles of int-c,
a variable competence of conferring a six-row phenotype to cc Vv
plants, with a single hex-v allele from 'Agneta' (TABLE IV). The
int-c allele from 'Agneta' is not among the "strong" alleles.
Nevertheless, this allele is likely to influence the phenotype
of 'Agneta' favourably. Mutants from the two-row state, at the
hex-v locus, have, generally, a variable development of lateral
florets, and in comparison with commercial six-row varieties
their lateral kernels are smaller, with shorter awns. An unin-
tentional selection for enhancing int genes occurring as spontaneous mutations seems likely in commercial six-row varieties.
The question arises which int loci have been the favoured ones
as enhancers.

The fact that the 'Agneta' c allele does not appear among
the "strong" alleles may have two completely differing meanings
for plant breeding. Either, an optimal gene co-operation has been
attained within the 'Agneta' variety, "stronger" c alleles running the risk of going beyond this optimum. Or, we are just in
the beginning of a deliberate utilization of the possibilities
inherent in the co-operation between int genes and the hex-v
locus.

CO-OPERATION BETWEEN INT GENES AND THE HEX-V GENE IN SINGLE DOSES

Most intermedium mutants are recessive; when they are crossed
to two-row barley, the F_1 does not deviate from the two-rowed
phenotype. The hex-v is semidominant, its F_1 with two-row barley
having the "six-row heterozygote"phenotype, which differs distinctly from both parents and has sterile lateral florets with
pointed lemmae. An interesting question arises whether an int
gene in heterozygous state can express itself phenotypically
against a "six-row heterozygote" background (i.e., in the genotype
Intint Vv) and, in such case, whether there are differences among
alleles and among different loci. In other words: Can a "flying
start" offered by the "six-row heterozygote" stimulate the heterozygous recessive int gene to add an effect?

F_1 data from crosses between single or double int mutants
and the hex-v allele hex 3 or 'Agneta' have been analysed. The
original spectrum of phenotypic denotations has been changed to
a system of numerical values. Transitions and demarcations may
be more or less artificial in this system and have to be regarded
with some caution. In a scale with 20 steps, the values 1-10
comprise "six-row heterozygous" phenotypes with pointed, sterile
lateral florets, or with gradually increasing awn development
and some kernel formation. The values 11-20 comprise "six-row-
resembling" phenotypes with increasingly pronounced kernel and
awn development and grain-filling of the lateral florets. It is
interesting that for the F_1 data derived from hex 3 there is a
sharper demarcation between the "six-row heterozygote" and the
"six-row-resembling" categories, the classes 8-10 failing here;
moreover, the classes 18-20 are also failing.

Data from F_1 plants with 2 or 3 heterozygous int loci are
entered in TABLE V. Our conclusions can be summarized as follows:

(i) Recessive int genes are able, in single doses already,
to add an effect if a promotion of lateral floret development
has been initiated by another gene.

(ii) This effect is reinforced when several such recessive
genes are brought together in single doses.

(iii) The amount of such positive co-operation depends on
the specific constellation of alleles belonging to different int
loci. Different alleles of hex-v also are probably of importance.

(iv) Different alleles of an int locus differ in their efficiencies to promote a six-row spike development.

TRIPLE MUTANT COMBINATIONS INT/INT/HEX-V

Combinations of double mutants int/int and the hexastichon
gene have produced beautiful six-row types with conspicuously big

spikes and thick straws; we have denoted them as "King-size".
Such types have not been observed before in crosses leading to
double mutants with int genes and the hex-v gene in various combinations.

TABLE VI shows frequencies with which "King-size" segregates
appear for the four most frequently mutated int loci (int-a, -c,
-d, and -e) when different int alleles are combined with other
int loci, and with the hex-v component either in the form of the
hexastichon mutant hex 3 or contributed by the commercial variety
'Agneta'. Both for the hex 3 and the 'Agneta' allele of hex-v
there appear very strong tendencies to heterogeneity among alleles of the individual int locus in their ability to produce the
"King-size" type, when they are combined with a second int locus.
The competence of a specific int allele is apparently dependent
on an interaction with the other loci in the constellation. There
are even indications that sometimes the "King-size" type does not
require strict triple homozygosity.

Since the 'Agneta' barley carries an int-c allele beside the
hex-v gene, an increased complexity must be taken into account
when an F_2 segregation involves four gene loci influencing the
spike development. On the other hand, new aspects may be opened
on the production of the "King-size" type when only int alleles
occupy the c locus in the segregating F_2 generation. From both
points of view, we should expect the increased production of
"King-size" plants as observed in the 'Agneta' part of TABLE VI.

Is the competence to produce the "King-size" phenotype once
and for all a given characteristic of the individual int allele,
or does it merely depend on an interaction among int loci? It is
obvious that the c allele of the 'Agneta' variety has, on the
whole, a poor competence. Locus int-h, with its 3 alleles, offers
another example of generally poor competence. On the other hand,
even if, on several occasions, a particular allele may rank among
the foremost ones when combined with two different loci, one,
nevertheless, finds positive correlation of only moderate, non-
significant size (TABLE VII). Thus, the interaction among loci
has a considerable influence on the degree of competence dis-
played by the particular allele.

There are interesting indications that the competence of int
genes to produce the "King-size" type has connections with their
efficiency to produce double mutants with regular six-rowed
spikes. Most successful in the latter respect are alleles belonging to int-c, int-d, and int-e. These three loci in constellations with one another and the hex-v also show the highest competence to produce the "King-size" phenotype. As an opposite extreme, int-h has only exceptionally produced the regular six-rowed
type in double mutant combinations and displays very poor competence for the production of "King-size" plants.

Step by step, it has become obvious that the triple mutant
phenotype "King-size" depends on particular kinds of gene interaction, particular int loci being more competent than others,
and the particular allele being competent depending on the constellation of loci. The "King-size" phenotype forms an illumina-
ting example of the importance of large-scale mutation research,

where a rich supply of mutants offers better possibilities to
find surprising and practically valuable forms of gene interaction.

PROGRAM FOR AN ANALYSIS OF THE INTERACTION BETWEEN INTERMEDIUM AND HEXASTICHON GENES

These instances of gene interaction promoting lateral floret
development in barley have prompted us to a systematical analysis,
beginning with simple gene systems and continuing with more complex systems, which will eventually culminate in the synthesis of
quadruple or quintuple mutant combinations of intermedium genes
alone or together with hex-v genes. The project comprises 8 subprojects, all of which were started in 1985. Three of the subprojects are investigations of the interaction among int loci,
from simple systems for a systematical adding of enhancing genes,
to more complex systems leading to triple or quadruple mutants,
respectively, segregating in the F_2 generation when double mutant
constellations of int genes are combined in crosses. The other
five subprojects (4, 5a, 5b, 6, and 7) are investigations of the
co-operation between int genes and the hexastichon gene hex-v
when, step by step, genes from 1, 2, 3, and 4 int loci are brought
together to enhance the hex-v gene.

Within such systems for analysis, we have aimed at the utilization of:

(a) mutants derived from the same mother variety;

(b) int loci co-operating to form regular six-rowed double
mutants;

(c) int alleles that are the most successful ones in this
respect.

The following characters are being studied, on the main
spike, on adequate numbers of plants: Awn development on lateral
florets relative to the main florets; fertility of lateral florets; kernel development of lateral florets; culm length to spike
basis; spike length.

Among the subprojects commented upon in the following, the
collection of data for nos. 1, 5a, and 5b has been finished, but
the statistical treatment remains.

1. Interaction among int loci, simple systems

The subproject aims at a comparison of the two-rowed mother
variety, single mutants, and double mutants with different heterozygous intermediaries obtained in the F_1 from adequate systems
of crosses. Which will be the effect of one of these loci in heterozygous state together with single alleles from one, two or
three other int loci? Which will be the effect of the same locus
in homozygous state, when it acts together with single alleles
from one or two of the other loci?

The study has been restricted to the 4 most frequently mutated int loci, int-a, -c, -d, and -e, involved in, totally, 20
systems of crosses, involving the alleles a 9, 10, 14, 27, and
30; c 5, 13, 29, 62, and 63; d 12, 22, 24, 28, and 36; and e 20,
23, and 65.

2. Interaction among int loci, more complex gene systems, leading to the synthesis of triple mutant F_2 segregates

The subproject aims at a comparison of the two-rowed mother variety, single mutants, and double mutants with different heterozygous intermediaries up to triple mutants obtained in the F_2 when double mutants with one of their int genes in common are combined in adequate crosses. Totally, 20 such crosses were started, each aiming at an F_2 segregation with 9 different genotypes where one homozygous int locus interacts with genes in two other int loci (0, 1, 2, 3, and 4 potential enhancers, respectively).

This analysis also is based on the four loci int-a, -c, -d, and -e, represented by 2, 2, 2, and 1 allele, respectively (a 9 and 27; c 62 and 63; d 12 and 36; and e 65) and combined in 17 different double mutant constellations.

3. Interaction among int loci, more complex gene systems, leading to the synthesis of quadruple mutant F_2 segregates

The subproject aims at a comparison of the two-rowed mother variety, single mutants, and double mutants with different heterozygous intermediaries up to quadruple mutants obtained in the F_2 when double mutants with none of their int genes in common are combined. This kind of cross involved the four genes a 9, c 63, d 12, and e 65 combined in the constellations (a 9 + c 63) x (d 12 + e 65). The quadruple mutant constellation is expected to segregate in the F_2 in the frequency 1/256.

4. Interaction between int loci and the hex-v locus, simple gene systems

The subproject aims at a comparison of two-rowed mother varieties, single int mutant, single mutant hex-v, and double mutant int/hex-v with different heterozygous intermediaries obtained in the F_2 when single int mutants are combined with six-row barley (the hexastichon mutant hex 3 or the commercial variety 'Agneta'). Totally, 24 such crosses were started, each aiming at an F_2 segregation with 9 different genotypes in a gene system comprising 0+0 up to 2+2 genes for the promotion of lateral floret development. The analysis is based on the four loci int-a, -c, -d, and -e, each represented by 3 alleles (a 9, 10, and 27; c 5, 13, and 29; d 12, 22, and 28; and e 20, 23, and 65).

5. Interaction between int loci and the hex-v locus, more complex gene systems, starting from triple mutants "King-size"

a. General analysis.- The subproject aims at a comparison of two-rowed mother varieties, single int mutants, double mutants int/int, single mutants hex-v, double mutants int/hex-v, and triple mutants int/int/hex-v ("King-size") with different heterozygous intermediaries obtained in the F_1 from adequate systems of crosses. Which will be the effect of the hex-v locus in heterozygous state when two int loci contribute with, altogether, 2, 3, or 4 genes for the promotion of lateral floret development? Which will be the effect of the hex-v locus in homozygous state when it acts together with the two int loci in heterozygous state?

The subproject is based on the 5 int loci -a, -c, -d, -e,

124

and -f (involving the 14 alleles a 27, 34, and 37; c 5, 13, 15, and 16; d 12, 24, 28, and 36; e 4 and 23; and f 19) in, totally, 20 systems of crosses.

b. F₂ segregation with 0, 1 or 2 hex-v genes against the background of two homozygous int loci.- The subproject aims at a study of double mutants int/int with 0, 1 or 2 hex-v alleles present for the promotion of lateral floret development. Totally, 16 double mutants int/int in combination with the hex-v allele hex 3 or the hex-v gene present in 'Agneta' have involved 8 alleles of int-a (a 1, 2, 8, 31, 32, 51, 52, and 61), 1 allele of int-b (b 6),7 alleles of int-c (c 5, 13, 15, 16, 29, 53, and 60), 3 alleles of int-d (d 24, 28, and 36), 2 alleles of int-e (e 4 and 23), and 1 allele of int-f (f 19).

6. Interaction between int loci and the hex-v locus, more complex gene systems, leading to the synthesis of quadruple mutant F₂ segregates

The subproject aims at a comparison of two-rowed mother varieties, double mutants int/hex-v, and triple mutants int/int/hex-v ("King-size") with different intermediaries up to quadruple mutants obtained in the F₂ when triple mutants int/int/hex-v with one of their int genes in common are combined. Totally, 28 such crosses were started, based on hex 3 or 'Agneta', and each leading to an F₂ segregation for the two int loci, against the background of two homozygous loci for the promotion of lateral floret development (int and hex-v). The basic level of 2+2 promoting genes will thus receive an extra contribution varying from 0 to 4 int alleles. The study is based on the four int loci int-a, -c, -d, and -e, represented by 5, 4, 4, and 2 alleles, respectively (a 1, 2, 27, 34, and 37; c 5, 7, 13, and 16, d 12, 24, 28, and 36; and e 4 and 23).

7. Interaction between int loci and the hex-v locus, more complex gene systems, leading to the synthesis of quintuplet mutant F₂ segregates

The subproject aims at a comparison of two-rowed mother varieties, six-rowed component hex-v, the relevant kinds of double mutants int/hex-v, and triple mutants int/int/hex-v with different intermediaries, up to quintuplet mutants obtained in the F₂ when triple mutants int/int/hex-v ("King-size") with none of their int genes in common are combined. Two such crosses were started, based on hex 3 and 'Agneta', and each leading to an F₂ segregation of four int loci against the background of the homozygous hex-v locus. The analysis involves the four int genes a 27, c 16, d 12, and e 23, combined in the constellations (a 27 + d 12) x (c 16 + e 23).

Acknowledgements.- I thank Arne Lundqvist for valuable discussions. This study was supported by grants from the Swedish Council for Forestry and Agricultural Research, from the Swedish Plant Breeding Board, and from the Erik Philip-Sörensen Foundation.

REFERENCES

(1) GUSTAFSSON, Å., HAGBERG, A., LUNDQVIST, U., PERSSON, G.,
A proposed system of symbols for the collection of barley
mutants at Svalöv, Hereditas 62 3 (1969) 409.

(2) NILAN, R.A., The cytology and genetics of barley. 1951-1962,
Res. Stud. Wash. State Univ. 32 1 (1964) 1.

(3) PERSSON, G., An attempt to find suitable genetic markers for
dense ear loci in barley, Hereditas 62 1 (1969) 25.

(4) GUSTAFSSON, Å., LUNDQVIST, U., Hexastichon and intermedium
mutants in barley, Hereditas 92 2 (1980) 229.

(5) LUNDQVIST, U., "Intermedium and hexastichon mutants in barley", Barley Genetics IV (Proc. 4th Int. Barley Genet. Symp.
Edinburgh, 1981) Edinburgh Univ. Press (1981) 908.

(6) LUNDQVIST, U., LUNDQVIST, A., "Barley mutants - diversity and
genetics", Barley Genetics V (Proc. 5th Int. Barley Genet.
Symp. Okayama,1986)(in press).

(7) GYMER, P.T., Probable allelism of Ii and int-c genes, Barley
Genet. Newsl. 7 (1977) 34.

(8) WOODWARD, R.W., The inheritance of fertility in the lateral
florets of four barley groups, Agron. J. 41 (1949) 317.

(9) ROBERTSON, D.W., Inheritance in barley, Genetics 18 (1933)
148.

(10) LEONARD, W.H., Inheritance of fertility in the lateral spikelets of barley, Genetics 27 (1942) 299.

(11) ROBERTSON, D.W., WIEBE, G.A., SHANDS, R.G., A summary of
linkage studies in barley: supplement II, 1947-1953, Agron.
J. 47 (1955) 418.

(12) GYMER, P.T., The genetics of the six-row/two-row characters,
Barley Genet. Newsl. 8 (1975) 44.

TABLE I. Comparison of mutagens for distribution of mutants on cer loci mutating with different frequencies

Mutagens compared	$\chi^2_{(7)}$
X rays vs neutrons	22.94**
Ethylene imine vs sulfonates	5.53 ns
EI + Sulf. vs other organics	3.52 ns
X rays vs all organics	30.16***
Neutrons vs all organics	82.35***
All organics vs NaN_3	16.85*
X rays vs NaN_3	28.19***
Neutrons vs NaN_3	42.90***

TABLE II. Means of spike phenotypes of double mutants in combinations of different int loci. In the lower half of the scheme, number of combinations.

	int-a	-b	-c	-d	-e	-f	-h	-i	-k	Mean
a		6.00	2.44	2.40	7.05	7.53	7.86	5.10	4.71	5.71
b	24		3.89	2.85	7.00		8.20	5.00	5.00	5.49
c	316	22		1.52	2.64	3.50	5.80	4.73	3.07	3.86
d	222	20	179		1.97	3.56	4.69	3.44	3.71	3.40
e	123	8	86	65		3.25	7.38	5.14	3.20	5.03
f	15		12	9	4		7.00	9.00	5.00	5.98
h	66	5	51	29	21	3		7.67	3.67	6.66
i	21	2	15	9	7	1	3		9.00	6.01
k	14	1	14	7	5	1	3	1		5.15

TABLE III. Spike development in double mutant combinations between int-a, -c, -d, and -e.

Double mutant	Numerical values of spike development									No. of comb.	Mean
	1	2	3	4	5	6	7	8	9		
a + c	170	38	3	22	77	2	2		2	316	2.44
d	116	31	4	16	52	2	1			222	2.40
e	5	1		1	37	5	12	1	61	123	7.05
c + d	146	11	1	6	14		1			179	1.52
e	41	15	1	2	21	4	1		1	86	2.64
d + e	41	10	1	2	10	1				65	1.97

TABLE IV. F$_2$ segregation for different int-c alleles crossed to 'Agneta'. Monohybrid segregation expected for hex-$\underline{v}$, without influence from the background of c genes.

int-c allele	F$_2$ six-row	Total	Chi-square analysis Six-row	χ²	df
5	68	255	(+) 0.28	0.37	
7	9	30	(+) 0.30	0.40	
13	77	267	(+) 1.57	2.09	
15	61	201	(+) 2.30	3.07	
16	70	272	(+) 0.06	0.08	
18	55	167	(+) 4.21	5.61*	
25	108	271	(+)23.91	31.88***	
33	8	62	(-) 3.63	4.84*	
38	39	169	(-) 0.25	0.33	
45	70	193	(+) 9.80	13.07***	
49	76	241	(+) 4.12	5.49*	
53	72	226	(+) 4.25	5.67*	
56	68	234	(+) 1.54	2.05	
60	13	38	(+) 1.29	1.72	
62	13	53	(+) 0.00	0.00	
63	39	107	(+) 5.61	7.48**	
76	47	186	(+) 0.01	0.01	
Total				84.16***	17
Pooled	893	2972	(+)30.28	40.37***	1
Heterogeneity				43.79***	16

TABLE V. F$_1$ spike phenotypes in numerical values when double mutants (of regular spike type) were combined with hex 3 and 'Agneta' barley.

int loci combined	No. of comb. with hex 3	No. of comb. with 'Agneta'	Mean hex 3	Mean 'Agneta'	Range hex 3	Range 'Agneta'
a + b	2	2	13.50	15.00	12 - 15	12 - 18
c	124	163	2.67	6.78	1 - 7	1 - 16
d	80	99	13.90	14.51	4 - 17	3 - 19
e	4	7	2.50	4.00	1 - 4	2 - 8
h	2	1	2.50	4	1 - 4	
b + c	4	4	3.00	7.75	1 - 4	4 - 11
d	11	11	13.18	15.83	6 - 17	11 - 20
e		1		7		
c + d	88	130	13.87	15.17	4 - 17	2 - 20
e	29	48	2.93	7.65	1 - 6	1 - 18
f	3	4	2.67	8.25	1 - 4	6 - 9
h	1	2	4	7.50		4 - 11
i	3	2	2.00	7.50	1 - 4	6 - 9
k	3	4	1.33	6.25	1 - 2	4 - 9
d + e	24	43	14.32	14.74	6 - 17	2 - 18
f	4	4	11.75	14.00	4 - 15	11 - 16
h	3	3	16.33	15.00	15 - 17	12 - 17
i	1	2	16	16.50		16 - 17
e + f	1	1	1	8		
h	1	1	1	6		
i	1		4			
k + k		2		7.00		6 - 8
h + k	1	1	1	8		6 - 8

TABLE VI. Frequencies of "King-size" plants in the F$_2$ from different combinations hex-v × int/int.

hex-v	int/int	Frequencies of "King-size" plants													χ² het
		0	.001-	.01-	.02-	.03-	.04-	.05-	.06-	.07-	.08-	.09-	.10-	.11	
hex 3	a / c		5	9	6	2				2	2	1			72.70***
	a / d		7	5	3	2				2	1				90.91***
	c / a		1	11	4	1									62.21***
	c / d		1	4	9	2	1								75.17***
	c / e		4	5	2				1			1	1		136.56***
	d / a		4	3		3	2								85.89***
	d / c		1	3	3	4			1						91.88***
	d / e		2	3	5	1									14.66 ns
'Agneta'	a / c		6	6	8	5	2	1							55.19***
	a / d	3	3	3	2	4	3	1		2	1				152.22***
	c / a		5	5	4	5	1	2							112.64***
	c / d			3	3	3	4	3	1		2	2			99.65***
	c / e	1		1	4	2	3	2							31.28**
	d / a		1	1	2	2	5								68.77***
	d / c		1	1	2	4	2			1	2		1		233.23***
	d / e	3		1	2		3	1							59.65***
	e / c	1	1	1	1	1	1								27.18***
	e / d	1		1	1	2		1							28.10***

TABLE VII. Comparison of F$_2$ frequencies of "King-size" plants for int-a alleles when (a + d) and (a + d) double mutants were crossed to 'Agneta' barley or hex 3.

int-a allele	'Agneta' (a + c)	'Agneta' (a + d)	hex 3 (a + c)
1	11/518	16/529	6/607
2	3/266	12/719	0/321
8	4/308	13/422	5/482
9	19/693	4/384	9/676
10	19/637	29/477	6/534
14	34/1080	14/831	19/1098
17	13/654	0/118	7/455
21	6/306	14/145	0/153
27	19/524	12/408	4/490
30	7/381	0/241	0/213
31	7/413		7/286
32	10/359	0/250	1/399
34	6/501	1/386	12/659
35	6/618	14/582	1/724
37	1/459	19/855	0/348
46	17/608	38/549	2/522
51	2/645	4/89	16/783
52	3/530	1/240	1/566
54	5/708	9/275	1/734
55	2/376		0/535
61	7/470	1/132	3/478
64	25/541	32/451	2/583

r (a+c)/(a+d) 'Agneta' = 0.305 ns

r (a+c)/(a+c) = 0.120 ns

BARLEY MUTANT HETEROSIS

M. MALUSZYNSKI*, A. FUGLEWICZ, I. SZAREJKO and A. MICKE*

Department of Genetics
University of Silesia
Katowice, Poland

* Joint FAO/IAEA Division
Plant Breeding and Genetics Section,
Vienna, Austria

M. Maluszynski (ed.), Current Options for Cereal Improvement, 129–146.
© *1989 by Kluwer Academic Publishers.*

ABSTRACT
     A high degree of heterosis was observed in F_1 of
spring barley induced mutant crosses. True breeding mutant
lines (M_7 or M_8) were used as parents in crosses among
each other or with the parent variety.   Heterosis of 20%
over the check variety was observed in 37 - 48% of analyzed
cross combinations, depending on investigated characters.
Agronomic performance or morphological characters of the
parent mutants cannot be used as a criterion to predict good
combining ability.   Tall or normal height mutants can give a
similar heterotic effect as semi-dwarf mutants.   Significant
heterosis was also observed in F_1 of crosses between
mutants and their parent variety. The paper summarizes the
results obtained during the last years with mutants from
cultivars Aramir, Diva, Karat, Salka, Plena and Trumph and
gives more data on the agrobotanical performance of the
mutants used as cross components.

INTRODUCTION

MNH (N-methyl-N-nitrosourea) should be considered as one of the most effective mutagens for induction of point mutations in cereals. After treatment with this mutagen (for methodology see Maluszynska and Maluszynski, 1983), we obtained more than three hundred dwarf or semi-dwarf barley mutants from different varieties. Many other morphological or physiological mutants with changes in grain, spike, leaf or tiller morphology, chlorophyll synthesis or improved powdery mildew resistance, were found as well and are now in our collection.

The design used to assess the heterosis effect in crosses of spring barley mutants with each other or with the parent cultivars has been described earlier by Maluszynski et al. (1988b). True breeding, mainly semi-dwarf or even dwarf mutant lines (M_7 to M_8) were used in these experiments. The potential of hybrids from these crosses has been evaluated by the level of heterosis over the best parent and over the check variety (original parent), space planted (24 x 25 cm). Because of lack of gametocides, all crosses were done by hand, which of course limited the number of F_1 plants for each combination to 20 or 30 plants in 4-5 replications.

PERFORMANCE OF SD-MUTANT HYBRIDS

Mutants from variety Aramir were mainly used to evaluate the usefulness of semi-dwarf induced mutants for hybrid barley. Mutant 239 AR gave a significant heterosis for grain yield in crosses with three other mutants from the same variety as well as with the parent (Table 1). The yield increase over the parent cultivar Aramir was always significant and ranged from 25.8 - 44.9%.

The mutant 239 AR is a semi-dwarf form with plant height of about 72% of the parent variety Aramir (89.4 cm $\pm$ 7.6). It produces only 4 to 5 tillers with a short first internode (10 cm). The lax spike (about 14 cm) is partly located inside the flag leaf sheath and leaf blades have about 20% larger surface than the parent Aramir. The yield of this late maturing mutant is usually 1/5 of that of the parent variety grown under the same conditions. The growth of the mutant is already delayed in the seedling stage. Seminal roots penetrate the soil only to 52% and the seedling height 10 days after sowing, reaches only 49% of the control plants.

The mutant 282 AR performs poorly as well. It is a form with bracteatum mutation, which means with a large extra basal glume-like bract in the collar region (similar form as described earlier by Gustafsson, 1947a). Semi- dwarf, about 78% of control plants' height, with short spike (8 cm) and short first internode (12 cm), this mutant produces small leaves with a surface of only 60% compared to the control. Poor tillering, small grain and a lower number of grain, resulted in a reduction of yield estimated by grain weight per plant at 37% of the parent variety.

The mutant 243 AR is also a semi-dwarf, with 59% the height of Aramir. It carries a mutation in the br locus responsible for the semi-dwarf character with a pleiotropic effect on leaves, spikes and awns. Very low tillering and other undesirable characters resulted in strong yield reduction. The other sd-mutant (280 AR) from variety Aramir, used as a cross component in this experiment, also performed poorly in comparison to Aramir variety.

Nevertheless, these mutants which were not at all attractive for plant breeders, when crossed with each other or with the parent variety, produced (F_1) hybrid plants surpassing the yield of the control by up to 25-44%. These crosses clearly demonstrated that the agrobotanical performance of the parent mutant, just like other inbred lines used for hybrids, cannot be used as a criterion for good combining ability.

The advantages for barley hybrid breeding of using semi-dwarf mutants developed from the same parent variety, with the high level of heterosis observed in crosses were discussed in a previous paper (Maluszynski et al., 1988a). Stiff-straw, semi-dwarf hybrids resistant to lodging under high level of fertilization are very attractive for barley production. For this reason the genetic analysis of semi-dwarfness in barley (Szarejko et al., 1988) has been

continued.   Up-to-date results of allelic tests of semi-dwarf
forms in our collection are presented in Table 2.
Preliminary investigations suggested that among investigated
mutants a minimum of 14 loci are responsible for this
character.   These results indicate as well that not only the
frequency of point mutations, responsible for dwarf or
semi-dwarf character, but also their diversity was very high
after mutagenic treatment of different barley varieties with
MNH.

OTHER MUTANTS FOR HYBRID PRODUCTION

     To determine whether the heterotic effect is connected
only with genes responsible for semi-dwarfism or is
independent from this character and related to mutant genetic
constitution, only tall and normal height mutants were chosen
for the next experiment.   Tall mutants (680 Q, 737 Q and 762
Q) as well as normal height mutants (670 Q, 796 Q and 800 Q)
originating from variety Karat (previously stock Q 448)
differed in their agrobotanical performance.   In our
experiments the mutants 680 Q, 737 Q, 762 Q and 796 Q were
superior to the parent in number and weight of grain per
plant.

     Also in this group of crosses, significant heterosis -
as compared to the better parent - was observed (Table 3).
The highest heterosis in weight of grain per plant (17.3g -
or an increase of 203.5%) in comparison to the parent variety
was noted in the cross of the tall mutant 762 Q with the
normal height mutant 796 Q.   Both of these mutants were
already superior to the parent in number and weight of
grain/plant, but even so, there was an 80% increase over the
better of the two mutant parents.

     No significant heterotic effect was observed regarding
the height of F_1 plants.   Relatively tall hybrids inherited
this character from tall mutants used as parents in crosses.

     Also three mutants derived from variety Diva showed
heterotic effects in intermutant or mutant/parent crosses.
Similar to semi-dwarf Aramir mutants the poor yielding mutant
116 DV gave statistically significant heterosis in two
crosses: with parent variety (80.0% in the weight of
grain/plant) and with another sd-mutant 125 DV (38.5%).   The
normal height mutant 134 DV expressed similar behaviour to
Karat mutants.   Although this mutant was superior to parent
variety Diva in such characters as the number of spikes and
grains/plant or grain weight/plant it gave as hybrid a
significant increase of all these characters, even if the
comparison was made with the better parent (Maluszynski et
al., 1988 b).

HETEROSIS IN CROSSES OF MUTANTS WITH PARENT VARIETIES

     The best examples of heterosis in grain yield, observed
in crosses of mutants with their original parent varieties
are summarized in Table 4.

The highest heterosis effect was observed in the combination Diva x 125 DV and Karat x 680 Q but as well in cross of variety Salka with its mutant 650 SL (a slightly taller but poorer yielding mutant). Important increases of yield were observed also in parent x mutant crosses with varieties Aramir and Plena.

We do not like to give the impression that all mutants or even all semi-dwarf mutants in crosses give a heterosis for grain yield in F_1. During our experiments (about 6 years) we described 18 out of 21 mutants from 4 varieties, which gave a significant effect of heterosis (20% or more over the better parent) in different crosses in relation to such important characters as, number of spikes per plant, number of grain per plant and weight of grain: In 48 cross combinations heterosis for grain yield (grain weight/plant) was observed in 23 cases (Table 5). Thus the probability to obtain higher yield from intermutant or parent/mutant crosses seems to be around 50%.

MUTANT HETEROSIS IS NOT CROP SPECIFIC

Mutant heterosis was already described in many papers for such species as Arabidopsis, barley, groundnut, maize, pea, pearl millet, Petunia axillaris, rice, sesame, sweet clover, and tomato.

For maize, Jones already reported in 1945 on heterosis resulting from back crosses of inbred mutant lines to the normal line. Mutants with so-called degenerative changes (like narrow leaf, dwarfness, crooked stalk, pale chlorophyll, blotched leaf and late flowering), which moderately reduced size and reproductive ability, showed heterotic effect in all cases when crossed with the parent. The percent of increase of F_1 over the higher yielding or taller parents ranged from 3 - 104% for grain yield and 3 - 9% for the height of plant. Narrow leaf and dwarf mutants expressed the highest heterosis in yield: doubled that of the parent line, while the height of these F_1 plants increased only 3 - 5%. It should be noted that such maize mutants with degenerative changes in crosses with unrelated lines did not reduce the yield of their offspring but even increased it. Monogenic inheritance observed for all these degenerative characters led the author to conclude that "heterosis could result from a single allelic difference if the change involved more than a single function".

Stoilov and Daskaloff (1976) found that induced maize mutants can give significant heterosis in yield in comparison to the standards. The authors confirmed results of other workers that induced mutations can change and often increase the general or specific combining ability of parental lines.

Pasztor et al. (1985) concluded on the basis of their research, that the net-assimilation rate has been improved in two- or three-line hybrids when maize mutants were used as parent lines. They believe that mutants are best used as parental lines in crosses producing three-line hybrids.

A maize inbred line was used by Dollinger (1985) for mutation induction by X-rays and EMS. He selected 31 visible recessive mutants for maize hybrid investigations. Recessive alleles in the homozygous stage showed numerous pleiotropic effects connected with plant vigour. He found that alleles with pleiotropic effects affected various vigour characteristics in the hybrid. The author concluded that "pleiotropic effects associated with recessive alleles provide a genetic basis for heterosis".

An example of the use of mutants to increase crop productivity in vegetative matter was given by Micke (1969, 1976) and Römer and Micke (1974). From a bitter variety of sweet clover (<u>Melilotus albus</u> Des.) 16 mutants with a low level of glucoside content were selected in M_2 and M_3 after mutagenic treatment. These mutants showed yield reduction but when crossed with each other in F_1 surpassed the parent variety in green matter production. Heterosis was observed in 73 out of 91 combinations. The author concluded that very effective mutagens such as X-rays and thermal neutrons produced so many changes in the treated genome that heterosis can be the result of multiple mutations.

Gorny (1976) investigated the heterosis effect in intermutant crosses of <u>Petunia axillaris</u> (Lam.) Juss. Mutants were obtained after MNH treatment of seeds from one self-pollinated plant of a 20 year old inbred line. A very strong heterotic effect was observed for such characters as height of plant, leaf and flower size, tillering or even the root system. This effect was very clear and significant, both in intermutant and mutant x parent crosses.

Qiu and Lu (1982) found in peanut, that heterosis from intermutant crosses was significantly higher than heterosis from crossing parental stocks. They concluded that two investigated mutants might be used as parental stocks for hybrid production, although their agronomic performance was poorer than the parental lines.

Induced semi-dwarf rice mutants showed heterotic effects similar to the barley mutants presented in this paper (Anandakumar and Sree Rangasamy, 1985). The TKM 6 mutant (about 87 cm height) was obtained from the tall (144 cm) variety TKM 6 . This mutant was analyzed in 17 crosses as a parent for the hybrid programme. The DGWG gene was present in various genetic backgrounds in parental lines used for crosses in this experiment. From data presented by the authors it is possible to conclude that the semi-dwarf character of the TKM 6 mutant is allelic to DGWG gene. Expression of heterosis was observed independent of a homo- or heterozygous stage of the DGWG gene in hybrids. The highest index value of heterosis was observed in crosses with the known semi-dwarf rice variety IR 8 carrying the DGWG gene. But almost the same heterosis was observed for crosses of this mutant with the tall, parent variety TKM 6 (index values 154.95; 143.82; 143.28 for IR 8 x TKM 6 mutant; TKM 6 x TKM 6 mutant; TKM 6 mutant x TKM 6, respectively). From the results of reciprocal crosses TKM 6 mutant x parent

variety one can get an idea of the complexity of the genetic
explanation of mutant heterosis. When the mutant was used as
female parent the heterosis for grain yield and the number of
productive tillers over TKM 6 reached 52.4% and 25.0%,
compared to 30.4% and 4.6% respectively when the mutant was
used as a pollinator.

Burton and Hanna (1977) described the heterotic effect
resulting from crossing specific, radiation induced pearl
millet mutants with their normal inbred parent. Some mutants
(necrotic, white-tipped and stubby headed) from Tift 23 when
crossed with the parent, significantly exceeded the yield of
standard hybrids.

Shumny (1977) has studied many F_1 mutant hybrids of
pea, tomato, corn and Arabidopsis in crosses with parent
lines and often observed a significant heterosis increase of
25 - 30% in comparison to the best initial form.

An extraordinarily high heterosis effect was found in
F_1 of fasciated pea mutants with other mutants or the
initial line (Gottschalk, 1976 a). In some crosses the
heterosis led to about 350% of the value of the initial line
in seed production. A clear heterotic effect was observed in
such characters as the total weight of green plant, pod and
seed number per plant. As an explanation of mutant heterosis
the author suggested heterozygosity for one single or a few
mutated genes but not the level of heterozygosity
characteristic for cross fertilizing species. The concept of
simultaneous mutations of closely linked genes was presented
in other papers (Gottschalk, 1968; Gottschalk 1976 b).

Similarly Blixt informed, during an FAO/IAEA Advisory
Group Meeting 1975 (Gottschalk, 1976 b), on an approximately
200% heterotic effect of pea in F_1 of mutant crosses in
comparison to the best market variety . Different chlorotica
mutants crossed with the parent variety or each other
expressed a high heterosis in seed yield per plant when grown
in well spaced conditions. Crosses of recessive or even
double recessive mutants were involved in these
investigations.

The heterosis effect caused by fasciated mutants of pea
was investigated also by Lönnig (1982). He suggested that
fasciata genes, recessive themselves, when homozygous can be
epistatic to several simultaneously mutated dominant genes
which are responsible for such characters as: lateness,
increased number and length of internodes or yield. Absence
of recessive epistasis and the resulting free expression of
dominant genes in F_1 may explain the heterosis effect in
F_1 hybrids. The occurrence of hybrid-like recombinants in
F_2 progenies supports the concept of hypostatic dominant
genes responsible for heterotic effect. This conclusion is
in contrast to a previously developed hypothesis that mutant
heterosis is due to overdominance of different fasciata
alleles in the heterozygous stage.

Similar to fasciata mutants in pea, Murty (1979) found
that F_4 mutants of sesame (Sesamum indicum L.) with a

different degree of fasciation in certain characters showed a significant heterotic effect in intermutant or mutant x parent variety crosses.

Kaul (1980) reported that hybrids of radiation induced early flowering mutants of garden pea exhibited a strong, statistically significant heterosis (26.5%) in total seed protein.

Heterosis in crosses of barley mutants was first found by Gustafsson (1946). Spontaneous chlorophyll mutants from the barley variety Golden were isolated in 1897. Two of these lethal mutants, albina 7 and xantha 3, showed a small heterotic effect in seed and spike number and grain weight per plant, when crossed with the normal type. This effect was significantly increased in so-called "dihybrid condition", which means in F_1 of these mutants (Gustafsson, 1947b). These early results were followed by investigations of other authors: Nybom (1950), Hagberg (1953), Doll (1966), Ellerström and Hagberg (1967), Gustafsson and Dormling (1972), Gustafsson et al. (1973 a,b,c), Mikaelsen (1973), Ahokas (1981), Maluszynski (1982), Fuglewicz and Maluszynski (1985). The problem of barley mutant heterosis was already discussed in our previous papers (Maluszynski et al., 1988 a,b).

MUTANT HETEROSIS AN OPEN INTERESTING PROBLEM
FOR RESEARCH AND PRAXIS

As mentioned above, heterosis has been observed in crosses of various morphological and physiological mutants of many different plant species. Heterosis can apparently derive from intermutant, mutant x parent, mutant x another variety or mutant x hybrid crosses. The expression of this effect can be different in reciprocal crosses.

The heterosis in mutant hybrids was observed independently from mutagens and methods applied for their induction. We found heterosis in about 50% of investigated mutants with no restriction to semi-dwarf or other specific morphological changes.

However it is possible to conclude from the large genetic diversity among mutants from the same variety, observed after using so-called "supermutagens", that a high number of different induced point mutations is to be expected in each cell nucleus. On this basis heterosis observed in crosses between mutants induced from the same parent variety could be explained as the result of multiple mutations, many of which are not recognized in mendelian segregation and have been called "background mutations".

Taking into consideration the presented examples and the various proposed explanations of mutant heterosis in different plant species, one may have to accept the possibility of different genetic interactions for particular species and even for particular mutants. The very large morphological and isozymatic variation of barley mutants observed in our experiments seems to support our conclusion (appearing also in other papers) - that the concept of simultaneous multiloci mutations is rather relevant.

Explanations of the heterosis phenomenon in general are not fully satisfactory, whether heterosis is derived from distant hybridization or from related mutant crosses. This should, however, not prevent plant breeders from considering its practical utilization.

The use of mutants with the same genetic background as cross-components for hybrid seed production has a number of advantages compared with the traditional method of distant hybridization. Mutants from the same parent variety may differ in several, but still, only in a limited number of genes. It can be expected that genes controlling characters important for F_2 seeds such as grain quality, cooking quality, taste and malting characteristics, will remain homozygous in F_1 and therefore will not segregate and will not cause any detectable heterogeneity in the harvested product.

Another specific advantage arises when semi-dwarf mutants of the same variety are used as cross components: even where superiority of F_1 mutant hybrids is strong in tillering and grain yield no complementation effect leading to tall plants will be observed in such cross combination if the mutant sd-alleles are the same. For hybrid seed production with the use of gametocides, parents with somewhat different plant height will be needed to facilitate an effective fertilization. This can also be achieved by the use of mutants from the same parent variety.

REFERENCES

AHOKAS, H., (1981) Re-evaluating the protein heterosis in
   hybrids of erectoids mutants in cv. Union. BGN, 11:
   21-23.

ANANDAKUMAR, C.R., SREE RANGASAMY,S.R., (1985) Heterosis
   and selection indices in rice. Egypt. J. Genet. Cytol.
   14: 123-132.

BURTON, G.W. and HANNA, W.W., (1977) Heterosis resulting
   from crossing specific radiation-induced pearl millet
   mutants with their normal inbred parent. Mutation
   Breeding Newsletter, 9: 3.

DOLL, H., (1966) Yield and variability of chlorophyll-
   mutant heterozygotes in barley. Hereditas, 56:
   255-276.

DOLLINGER, E.J., (1985) Effects of visible recessive
   alleles on vigor characteristics in a maize hybrid.
   Crop Science, 25: 819-821.

ELLERSTRÖM, S and HAGBERG, A., (1967) Monofactorial
   heterosis in autotetraploid barley. Hereditas, 57:
   319-326.

FUGLEWICZ, A. and MALUSZYNSKI, M., (1985) Heterosis in
   hybrids of spring barley mutants. BGN. 15: 30-32.

GORNY, A., (1976) Chemomutanty Petunia axillaris. PhD.
   Thesis, Silesian University, Katowice.

GOTTSCHALK, W., (1968) Simultaneous mutation of closely
   linked genes. A contribution to the interpretation of
   'pleiotropic' gene action. In: Mutations in plant
   breeding II, IAEA, Vienna, 97-109.

GOTTSCHALK, W., (1976a) Pleiotropy and close linkage of
   mutated genes. New examples of mutations of closely
   linked genes. In: Induced mutations in cross
   breeding, IAEA, Vienna, 71-78.

GOTTSCHALK, W., (1976b) Monogenic heterosis. In: Induced
   mutations in cross-breeding. IAEA, Vienna, 189-197.

GUSTAFSSON, A., (1946) The effect of heterozygosity on
   variability and vigour. Hereditas, 32: 263-286.

GUSTAFSSON, A., (1947a) Mutations in agricultural plants.
   Hereditas, 33: 1-99.

GUSTAFFSON, A., (1947b) The advantageous effect of
   deleterious mutations. Hereditas, 33: 573-575.

GUSTAFSSON, A. and DORMLING, I., (1972) Dominance and overdominance in phytotron analysis of monohybrid barley. Hereditas, 70: 185-216.

GUSTAFSSON, A., EKMAN, G. and DORMLING, I., (1973a) Vigour and variability of phytotron-cultivated monohybrid barley. Additional information. Hereditas, 73: 1-10.

GUSTAFSSON, A., DORMLING, I. and EKMAN, G., (1973b) Phytotron ecology of mutant genes. I. Heterosis in mutant crossings of barley. Hereditas, 74: 119-126.

GUSTAFSSON, A., DORMLING, I. and EKMAN, G., (1973c) Phytotron ecology of mutant genes. II. Dynamics of heterosis in an intralocus mutant hybrid of barley. Hereditas, 74: 247-258.

HAGBERG, A., (1953) Heterozygosity in erectoides mutations in barley. Hereditas, 39: 161-178.

JONES, D.F., (1945) Heterosis resulting from degenerative changes. Genetics, 30: 527-542.

KAUL, M.L.H., (1980) Radiation genetic studies in garden pea. 9.Non-allelism of early flowering mutants and heterosis. Z. Pflanzenzuchtg. 84: 192-200.

LÖNNING, W.E., (1982) Dominance, overdominance and epistasis in Pisum sativum L. Theor. Appl. Genet. 63: 255-264.

MALUSZYNSKI, M., (1982) The high mutagenic effectiveness of MNUA in inducing a diversity of dwarf and semi-dwarf forms of spring barley. Acta Societatis Botanicorum Poloniae, 51: 429-440.

MALUSZYNSKA, J. and MALUSZYNSKI, M. (1983) MNUA and MH mutagenic effect after double treatment of barley seeds in different germination periods. Acta Biologica, Katowice, 11: 238-248.

MALUSZYNSKI, M., MICKE, A., SIGURBJOERNSSON, B., SZAREJKO, I. and FUGLEWICZ, A., (1988a) The use of mutants for breeding and for hybrid barley. In: Barley Genetics V, Proc. Fifth Intern. Barley Genet. Symp., Okayama (in press).

MALUSZYNSKI, M., SZAREJKO, I., MADAJEWSKI, R., FUGLEWICZ, A. and KUCHARSKA, M., (1988b) Semi-dwarf mutants and heterosis in barley. I. The use of barley sd-mutants for hybrid breeding. In: Semi-dwarf cereal mutants and their use in cross breeding III, IAEA, Vienna, TECDOC-455: 193-206.

MICKE, A., (1969) Improvement of low yielding sweet clover mutants by heterosis breeding. In: Induced mutations in plants, IAEA, Vienna, 541-550.

MICKE, A., (1976) Hybrid vigour in mutant crosses.
Prospects and problems of exploitation studied with
mutants of sweet clover. In: Induced mutations in
cross-breeding, IAEA, Vienna, 199-218.

MIKAELSEN, K., (1973) Studies on the inheritance of the
high seed protein content of an erectoides mutant
(H-14) in barley. In: Nuclear techniques for seed
protein improvement. IAEA, Vienna, 217.

MURTY, G.S.S., (1979) Heterosis in inter-mutant hybrids
of Sesamum indicum L. Curr. Sci. 48: 825-827.

NYBOM, N., (1950) Studies on mutations in barley
I. Superdominant factors for internode length.
Hereditas 36: 321-328.

PASZTOR, K., PEPO, P. and EGRI, K. (1985) Changes in the
production of maize hybrids due to mutant parent
lines. Acta Agronomica Academiae Scientiarum
Hungaricae, 34: 189-195.

QIU Qing-shu and LU Rong-rong (1982) Heterosis of F_1
hybrids of peanut mutants and F_1 hybrids of original
stocks. Application of Atomic Energy in Agriculture,
Beijing, 1: 13-14.

RÖMER, F.W., MICKE, A., (1974) Combining ability and
heterosis of radiation-induced mutants of Melilotus
albus DES. In: Polyploidy and induced mutations in
plant breeding, IAEA, Vienna, 275-276.

SHUMNY, V.K. (1977) Using mutants for obtaining
heterosis. Mutation Breeding Newsletter, 9: 3.

STOILOV, M. and DASKALOFF, S., (1976) Some results on the
combined use of induced mutations and heterosis
breeding. In: Induced mutations in cross-breeding,
IAEA, Vienna, 173-188.

SZAREJKO, I., MALUSZYNSKI, M., NAWROT, M. and SKAWINSKA-
ZYDRON, G. (1988) Semi-dwarf mutants and heterosis in
barley. II. Interaction between several mutant genes
responsible for dwarfism in barley. IN: Semi-dwarf
cereal mutants and their use in cross breeding III.
IAEA, Vienna, TECDOC-455: 241-246.

TABLE 1: Heterosis manifestations (% increase of F_1 over the better parent \<a\> and original variety \<b\>) in selected crosses with the mutant 239AR (Maluszynski et al., 1988b - modified)

Cross	Height			No. of spikes per Plant		Grain per plant			
	(cm)		(%)	(No.)	(%)	Number		Weight	
						(No.)	(%)	g	(%)
Aramir x 239 AR	90.8	a	1.6	16.6	1.2	362.4	32.3*	11.2	25.8*
		b	1.6		1.2		32.3*		25.8*
280 AR x 239 AR	94.4	a	7.4	18.4	16.4	374.4	57.7*	12.9	86.9*
		b	5.6		12.2		36.7*		44.9*
282 AR x 239 AR	91.8	a	30.9*	18.8	62.1*	403.2	171.5*	12.0	263.6*
		b	2.7		14.6		47.2*		34.9*
243 AR x 239 AR	90.9	a	34.4*	17.5	216.4*	378.8	297.6*	12.3	485.7*
		b	1.7		6.5		38.3*		38.2*
Aramir	89.4 ± 7.6			16.4 ± 6.5		273.9 ± 98.4		8.9 ± 3.7	

* $p < 0.05$

TABLE 2: Allelic relationship of semi-dwarf and dwarf spring barley mutants selected from our collection

Parent variety	No. of dwarf or semi-dwarf mutants	No. of investigated mutants	No. of $\underline{sd}$ or $\underline{dw}$ loci	No. of allelic mutants in locus min.	max.
Julia	39	20	8	1	11
Delisa	42	28	6	3	14
Plena	8	8	2	1	7

144

TABLE 3: Heterosis manifestations (% increase over the better parent <a> or original <b>) in F_1 of selected Karat's mutant crosses (Maluszynski et al., 1988 b - modified) (data from 1985)

Cross	Height (cm)			No. of spikes per Plant (No.)	(%)	Grains per plant Number (No.)	(%)	Weight g	(%)
Karat x 680 ♀ (=♀ 448)	90.4	a	2.4	16.0	58.4*	333.8	65.6*	12.0	93.5*
		b	12.1*		68.4*		68.1*		110.5*
737 ♀ x 670 ♀	99.2	a	0.0	16.0	48.1*	388.9	69.1*	15.5	106.7*
		b	23.1*		58.4*		92.9*		171.9*
737 ♀ x 796 ♀	102.7	a	3.5	16.3	29.4*	365.4	50.9*	13.7	71.3*
		b	27.4*		61.4*		81.3*		140.4*
762 ♀ x 796 ♀	90.0	a	2.2	18.6	44.2*	385.0	45.7*	17.3	84.0*
		b	11.7*		84.2*		91.4*		203.5*
800 ♀ x 737 ♀	98.1	a	-1.1	17.1	58.3*	393.3	71.0*	14.2	89.3*
		b	21.7*		69.3*		95.1*		149.1*

*/ $p < 0.05$
normal height - 670♀, 796♀, 800♀
tall mutants - 680♀, 737♀, 762♀

TABLE 4: Heterosis (% increase) observed in crosses of original varieties with their mutants in comparison to the better parent <a> and original variety <b>

Cross		Height	Spikes No./Plant	Grain per plant Number	Weight
Aramir x 239 AR	a	1.6	1.2	32.3*	25.8*
	b	1.6	1.2	32.3*	25.8*
Salka x 650 SL	a	16.2**	26.1**	63.9**	79.6**
	b	16.2**	26.1**	63.9**	79.6**
Karat x 680 Q	a	2.4	58.4**	65.6**	93.5**
	b	12.1*	68.4**	68.1**	110.5**
Plena x 841 Pl	a	-2.9	25.5	39.0*	43.4*
	b	-2.9	25.5	39.0*	43.4*
Diva x 125 DV	a	11.5**	62.7**	87.6**	115.3*
	b	11.5**	62.7**	87.6**	115.3*

* $p < 0.05$
** $p < 0.01$

TABLE 5: Number of combinations with heterotic effect over 20%

Variety	No. of mutants used for crosses	No. of mutants as parents with heterotic effect	No. of cross combinations	No. of combinations with heterotic effect in relation to different characters			
				Height	No. of spikes per plant	No. of grain per plant	Grain weight per plant
Trumph	5	3	10	0	3	2	2
Diva	3	3	6	1	4	5	5
Aramir	6	5	15	0	1	5	5
Karat	7	7	17	1	10	12	11
TOTAL	21	18	48	2	18	24	23

RICE IMPROVEMENT (INVOLVING ALTERED FLOWER STRUCTURE MORE SUITABLE TO CROSS-POLLINATION) USING IN VITRO TECHNIQUES IN COMBINATION WITH MUTAGENESIS

MIN SHAOKAI, WANG CAILIAN*, WANG GUOLIANG,
XIONG ZHENMIN, QI XIUFANG

China National Rice Research Institute (CNRRI)
Hangzhou, China, P.R.

*Institute of Utilization of Atomic Energy
Zhejiang Academy of Agricultural Sciences
Hangzhou, China, P.R.

M. Maluszynski (ed.), Current Options for Cereal Improvement, 147–152.
© *1989 by Kluwer Academic Publishers.*

148

ABSTRACT
It has been proven that irradiation treatment and in vitro culture are effective means of rice improvement. Recent advances in using the integrated method (combination of in vitro culture and mutagenesis) show more efficiency in plantlets regnerated and variants induced. In China there is great potential to increase the yield of hybrid rice seed production through alteration of the flower structure of three lines more suitable to cross-pollination, based on the availability of broad genetic variation for floral traits in rice germplasm resources. The design of research project on rice improvement (involving altered flower structure more suitable to cross-pollination) using in vitro techniques in combination with mutagenesis has thus been determined.

INTRODUCTION

Since Ichijima's report of mutations induced by x-irradiation in rice [1], many rice breeders have shown an interest in mutation breeding. In China, up to now, 65 varieties have been developed through mutation breeding, five of which have reached over 100,000 ha extension area. The first mutation variety is Ai-Fu 9 (1964) and the variety with the largest extension area is Yuan-Feng-Zao (1 million ha).

DH IN RICE BREEDING

For anther culture, since 1964 when Guha and Maheshwari obtained haploid plants from anther culture of Datura, haploid plants have been obtained from 206 species of 70 genus. The anther culture method started in 1970 in China, and 15 rice varieties have been developed by the haploid techniques, such as japonica type Zhe-Keng 66 (6,700 ha in 1985), Zhong-Hua 8 and 9 (10,000 ha in 1985), etc. Up to now the rate of pollen-derived green plants on the basis of anther number was 0.5% for indica rice and 5-6% for japonica rice. So haploid breeding through anther culture has been available in japonica rice, but there are still hindrances for using this approach in indica rice. In order to increase the culture ability of indica rice, screening of genotypes, improving inoculation methods and culture conditions such as light and pretreatment temperature of donor plants are underway at CNRRI.

The potential usefulness of somaclonal variation for plant improvement first became apparent in sugarcane. In the Hawaiian Sugar Planters' Association Experimental Station's annual reports of 1967, 1968 and 1970, mention is made of variability among plants derived from sugarcane tissue cultures. In 1978, Oono reported that the variation frequency of the plants regenerated from rice seeds could reach 71.9%.

In 1979 we started research on somaclonal variation by using young rice panicles, and matured embryos. Up to now, research on somatic genetics has developed as an active branch in the realm of plant genetics and breeding. Somaclonal variation has such advantages as high frequency, wide range, and breeding true in the first self-pollinated progeny [4]. T42, a new somaclonal line developed by CNRRI covered an area of 1,500 ha in 1985. In addition to the somaclonal variations occuring in the process of cell and tissue culture, a certain tendency for oriented selection of mutated cells is also included in the culture process. These research activities have day by day manifested their potential application for varietal improvement.

IMPROVING THE FLORAL STRUCTURE

Rice is a strictly self-pollinating crop. The three lines used in the hybrid rice system at present still keep the flower structure proper for selfing. With artificial supplement, the seed-set percentage of outcrossing of existing male sterile lines in the field is generally 20-30% for _indica_ type and 40-50% for _japonica_ type. It is estimated that when the seed set percentage is increased by 1%, the yield of hybrid seed production will be increased by 44-94 kg per ha. So, there is great potential to increase the yield of seed production of hybrid rice by improving the floral traits of three lines to make them more suitable for cross-pollination.

In general, male sterile lines with big and exerted stigma, long duration of anthesis, or the maintainer line and restorer line with large anthers, large quantity of pollen and good anther dehiscence are expected. Quite a number of scientists have reported in detail that the anther size, amount of pollen, stigma size and rate of stigma exertion of some wild rice species are obviously longer or higher than those of cultivated rice respectively [5,6,7,8,9]. Within cultivated rice, there also exists somewhat wider variations for flower structure [10]. All of these provide rich germplasm resources for improving the floral structure.

In China, a number of new lines which have floral structure somewhat suitable for outcrossing have been developed through hybridization between wild rice and cultivated varieties [11,12]. With this method the process is slow and new lines are now and then of undesirable linkage characters or of poor disease resistance and thus cannot meet practical needs. Therefore, it is necessary that integrated breeding technology be adopted for further development of the new line with improved floral structure suitable for outcrossing.

It has been proven that wide variation will appear in the progeny population through both mutagen treatment and in vitro culture, and that anther culture can shorten the breeding cycle. If these methods are adopted integrately, good results will be obtained in rice improvement (involving altered flower structure more suitable to cross-pollination). Chinese scientists have conducted investigations on the effectiveness of combining in vitro culture and mutagenesis. It has been pointed out that through a combination of lower dosage or irradiation and anther culture, both the frequency of callus induction and green plantlet regeneration can be raised [13]. Besides, it has been found that in anther culture [14] and somatic tissue culture [15], treatment of radiation can stimulate wider variation in progenies.

REFERENCES

[1]   ICHIJIMA, K. (1934)   On the artificially induced mutations and polyploid plants of rice occuring in subsequent generations.   Proc. Imp. Acad. Japan (Tokyo), 10: 388.

[2]   GUHA, S. and MAHESHWARI, S.C. (1964)   In vitro producing of embryo from anthers of Datura. Nature, 204: 497.

[3]   OONO, K. (1978)   Test tube breeding of rice by tissue culture. Trop.Agric. Res. 11: 109-124.

[4]   SUN ZONGXIU, ZHAO CHENGZHANG, ZHENG KANGLE, QI XIUFAN and FU YAPING   (1983)   Somaclonal genetics of rice. TAG. 67: 67-73.

[5]   SAMPATH, S. (1962)   The genus Oryza: Its taxonomy and species interrelationship.   Oryza, 1: 1-29.

[6]   OKA, H.I. and MARISHIMA, H. (1967)   Variation in the breeding system of wild rice Oryza perrennis. Evolution, 21: 249-258.

[7]   VIRMANI, S.S. and ATHWAL, D.S. (1973)   Genetic variability in floral characteristics influencing outcrossing in Oryza sativa L. Crop Sci. 13: 66-67.

[8]   PARMAR, K.S., SIDDY, E.A. and SWAMINATHAN, M.S. (1979) Variation in anther and stigma characteristics in rice.   Indian Journal of Genetics and Plant Breeding, 39: 551-559.

[9]   LI QINXIU, LIU BIAOXI and WANG YULAN   (1981)   Studies of O. longistaminata and its application.   Sichuan Agricultural Sciences and Technology, 6: 10-20 (in Chinese).

10]   XU YUNBI, SHEN ZONGTAN, YANG ZAINENG and YING CUNSHAN (1986)   Study of improving percentage of cross-pollination in rice, 1. Analysis on variation of stigma exertion in O. sativa L.   Acta Agriculturea Universitatis Zhejiangensis, 12 (4): 359-368 (in Chinese, English summary).

[11] LI QINXIU, LIU BIAOXI and WANG YULAN (1981) Development of long-stigma rice male sterile line. Sichuan Agricultural Sciences and Technology, 2: 13-14 (in Chinese).

[12] YANG RENCUI and LU HAORAN (1986).   Study on combining ability in floral characters of rice.   Paper of International Symposium on Hybrid Rice, Changsha, China.

[13] YIN DAOCHUAN, WEI QIJIAN, YU QIUCHENG and WANG LU
    (1982) Study of the effect of radiation on rice
    anther culture. Application of Atomic Enregy in
    Agriculture, 1: 28-33 (in Chinese).

[14] HUANG SHIZHOU, HU CHUNG and CHUENG CHENGCHI (1977)
    Effects of gamma radiation pretreatment on
    induction of rice pollen plants. Reports of the
    Academic Seminar on Anther Culture, 279-280 (in
    Chinese).

[15] ZHAO CHENGZHANG, SUN ZONGXIU, QI XIUFANG, ZHENG KANGLE
    and FU YAPING (1983) The effects of ^{60}Co gamma
    radiation on induction of rice regenerated plants
    and their traits. Journal of Cytobiology, 1:
    21-24 (in Chinese).

USE OF DOUBLED-HAPLOID LINES IN RECURRENT AND NATURAL SELECTION IN BARLEY

E. REINBERGS, D.E. MATHER AND J.D. PATEL

Crop Science Department
University of Guelph
Canada

M. Maluszynski (ed.), Current Options for Cereal Improvement, 153–160.
© *1989 by Kluwer Academic Publishers.*

154

ABSTRACT

Doubled-haploid lines are employed in two barley (Hordeum vulgare L.) population improvement efforts: a diallel recurrent selection program and a natural selection program.

In each cycle of the recurrent selection program, doubled-haploid lines are derived from a diallel cross. These lines are evaluated in hill plots to select superior lines to be inter-crossed for the next cycle. In the initial cycle (Cycle 0) of this program, selection was effective in improving grain yield per hill, number of spikes per hill, and plant height. Selection among Cycle 1 lines emphasized grain yield per hill, grain yield per spike, and number of spikes per hill. In Cycle 2, selection will emphasize grain yield, with some attention paid to maintaining contributions from each of the original parents, to avoid excessive narrowing of the germplasm base.

Natural selection has been studied in two populations originating from the same diallel: a mixture of doubled-haploid lines and a segregating composite cross. Natural selection favoured higher grain yield, later heading, and taller plant stature. Alternation of these populations between two dissimilar locations was detrimental to yield performance, and did not enhance adaptability or variability. Composite-cross populations, which had potential for segregation, recombination, and heterosis during natural selection, had higher mean grain yields and headed later than homozygous doubled-haploid mixtures.

1. INTRODUCTION

Doubled-haploid (DH) lines can be useful tools in plant breeding, and in studies of plant breeding methods. At the University of Guelph, DH barley (Hordeum vulgare L.) lines, produced using the bulbosum method [1], are employed in a diallel recurrent selection scheme, and in an investigation of the effects of natural selection. This paper describes these two population improvement programs and reviews their progress.

2. RECURRENT SELECTION PROGRAM

Recurrent selection employs repeated selection and inter-crossing to increase the frequency of favourable alleles in breeding populations. It should be more efficient than simple hybridization for accumulating favourable alleles affecting quantitatively inherited traits. Formal recurrent selection schemes have not been extensively used to improve self-pollinating

crops. In such crops, it may be desirable to select among homozygous progenies. However, selfing several generations after
intercrossing would reduce the efficiency of recurrent selection
by increasing the time required to complete each cycle.

Incorporation of doubled-haploidy into recurrent selection
schemes can introduce efficiency due to selection among homozygous
lines [2] without extending the time required per cycle. Choo et
al. [3] proposed a population improvement scheme for self-
pollinating crops, which should require only two to three years
per cycle. It involves intercrossing lines in a diallel or partial
diallel, extracting 20 DH lines from each cross, and conducting
yield tests to select lines to be intercrossed in the following
cycle.

We have used this DH diallel recurrent selection plan for
population improvement of six-row barley in Ontario, Canada. In
each cycle, seven lines are intercrossed in diallel. The bulbosum
method [1] is used to derive 20 DH lines from each cross. After
off-season seed increase, the DH lines are yield-tested in hill
plots at two locations. Seven superior entries are selected and
intercrossed to initiate the next cycle. Figure 1 illustrates this
plan.

The initial (Cycle 0) parental lines included four cultivars
and three breeding lines of six-row barley. Five of these parents
originated from breeding programs in central Canada (Ontario and
Quebec), one (Paragon) was from Manitoba, and one (64-76, a semi-
dwarf) was from Minnesota. These two introduced lines are
relatively low-yielding under Ontario conditions (Figure 2).

In 1976, the seven Cycle 0 parents were diallel-mated,
ignoring reciprocals. Twenty DH lines were produced from 17 of the
resulting 21 F_1s. Because of difficulties encountered in deriving
doubled haploids from some crosses, less than 20 lines were
produced from the remaining four crosses. A total of 398 Cycle 0
DH lines were tested in hill plots in 1978. Grain yield, number of
spikes per hill, 1000-grain weight, plant height, heading date,
and maturity date were recorded. Seven Cycle 1 parents were
selected based on these characters.

These selections were intercrossed; 260 Cycle 1 DH lines were
produced and were tested in hill plots in 1982. Grain yield per
hill, number of spikes per hill, and plant height responded to
selection in the first cycle [4]. Relatively high heritability
values were estimated for grain yield per hill, grain yield per
spike, and plant height. Selection of the Cycle 2 parents
emphasized grain yield per hill, grain yield per spike, and number
of spikes per hill. Lines originating from poor crosses were not
selected to be used as parents.

In Cycle 2, 350 DH lines were produced. They were tested in
1986. Data has been recorded on grain yield per hill, heading
date, plant height, and mildew resistance. We expect to observe
further progress for grain yield per hill. Once the data from the
1986 test has been processed, seven parents will be selected for

Cycle 3. Selection will emphasize grain yield. Some attention will be paid to maintaining contributions from each of the original parents. In previous cycles, selection of more than one line from any individual cross was avoided. Nevertheless, examination of the ancestry of the material in the current cycle (Table I) provides cause for concern that progeny of certain parents could be eliminated from the population.

Until data on the second cycle of selection is analyzed, conclusions regarding progress from recurrent selection can not be drawn. To date, no cultivars have been released from this program. However, one of the Cycle 1 lines is currently in advanced testing, and several Cycle 2 lines will be entered in further yield trials in 1987. Some Cycle 2 crosses show promise as sources of resistance to current races of powdery mildew.

The objective of producing 420 DH lines (20 from each of 21 crosses) was not met in Cycles 0, 1 or 2. Genotypic variability for responsiveness to the bulbosum technique may account for the difficulty experienced in deriving doubled-haploid lines from certain crosses. In Cycle 2 we attempted to overcome these difficulties by including reciprocal crosses, and by using a mixture of pollen from several H. bulbosum clones.

The duration of each complete cycle has been four years, rather than the two to three years proposed by Choo et al. [3] because of: (1) efforts to obtain a reasonably balanced representation of the diallel despite the difficulty encountered in obtaining DH lines from certain crosses and (2) the time required to process material and data from yield trials which included as many as 14,000 hill plots per cycle.

Data from this program have been subjected to genetic analysis [5]. In conducting this recurrent selection program, emphasis has been placed upon both breeding and research objectives, so that contributions might be made to quantitative genetics theory, and to the understanding of barley genetics, as well as to barley improvement.

3. NATURAL SELECTION PROGRAM

Natural selection can effect changes in breeding populations which are advanced in bulk. The genetic architecture of a population, and the nature of the environment in which it is grown, may be expected to affect the direction and magnitude of these changes.

We have studied the effects of natural selection in different environments on two populations of common ancestry but contrasting genetic structure.

Both populations originated from the Cycle 0 diallel described above; one was a mixture 398 Cycle 0 DH lines and the other was a composite of F_3 progeny from the 21 Cycle 0 crosses. In the composite cross, generation of new genotypes through segregation and recombination would be possible, but in the initially homozygous DH mixture, only the frequencies of component genotypes could change.

These populations were advanced under natural selection in two dissimilar locations, and with alternation between locations. After five years of natural selection, random lines from each population were evaluated. The details of this experiment are being published elsewhere [6].

Natural selection generally favoured higher grain yield, later heading, and taller plant stature. Composite-cross populations, had higher mean grain yields and headed later than homozygous DH mixtures. This suggests that they benefited from segregation and recombination and/or from heterosis. While the frequency of high-yielding lines was higher in the composite cross population, the best DH line was just as high-yielding as the best composite cross line.

The purpose of alternating populations between locations was to test the hypothesis, suggested by previous research [7;8], that alternation between contrasting environments might broaden adaptation. Our results indicated that the alternation treatment was detrimental to yield performance, and did not enhance adaptability or variability.

ACKNOWLEDGEMENTS

The research and breeding programs reviewed in this paper are long-term efforts, initiated in 1976, and ongoing today. The authors would like to acknowledge the involvement of Drs. T.M. Choo, S.O. Fejer, A. Kotecha, and J.D.E. Sterling, and the technical assistance of L. Christie, M. El Halwegy, M. Etienne, L. Taurins, and M. White.

REFERENCES

[1] KASHA, K.J., KAO, K.N., High frequency haploid production in barley (Hordeum vulgare L.), Nature 255 (1970) 874.
[2] GRIFFING, B., Efficiency changes due to use of doubled-haploids in recurrent selection methods, Theor. Appl. Genet. 46 (1975) 367.
[3] CHOO, T.M., CHRISTIE, B.R. REINBERGS, E., Doubled haploids for estimating genetic variances and a scheme for population improvement in self-pollinating crops, Theor. Appl. Genet. 54 (1979) 267.
[4] PATEL, J.D., REINBERGS, E., FEJER, S.O., Recurrent selection in doubled-haploid populations of barley (Hordeum vulgare L.), Can. J. Genet. Cytol. 27 (1985) 172.

[5] CHOO, T.M., KOTECHA, A., REINBERGS, E., SONG, L.S.P., FEJER, S.O., Diallel analysis of grain yield in barley using doubled-haploid lines, Plant Breeding 97 (1986) 129.

[6] PATEL, J.D., REINBERGS, E., MATHER, D.E., STERLING, J.D.E. AND CHOO, T.M., Effects of natural selection in a doubled-haploid mixture and a composite cross of barley, Crop Sci. (1987, in press)

[7] CHOO, T.M., KLINCK, H.R., ST. PIERRE, C.A., The effect of location on natural selection in bulk populations of barley (Hordeum vulgare L.) II. Quantitative traits, Can. J. Plant Sci. 60 (1980) 41.

[8] ST. PIERRE, C.A., H.R. KLINCK, GAUTHIER, F.M., Early generation selection under different environments as it influences adaptation of barley, Can. J. Plant Sci. 47 (1967) 505.

TABLE I. Contributions of Original (Cycle 0) Parental Lines to Cycle 2 Population in Barley Doubled-Haploid Diallel Recurrent Selection.

Cycle 0 Parent	Coefficient of Relationship with Cycle 2 population
Paragon	0.214
64-76	0.357
Trent	0.214
OB114-4	0.036
Vanier	0.071
OB128-10	0.036
Champlain	0.071

160

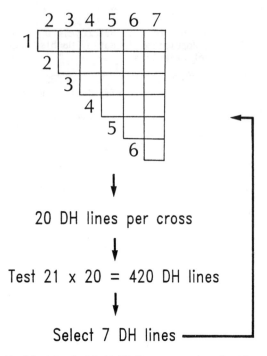

FIG. 1.   *Doubled haploid diallel recurrent selection scheme*

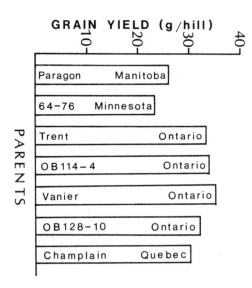

FIG. 2.   *Grain yield of seven Cycle 0 parental lines of barley,*
evaluated in hill plots at two locations in Ontario in
1978.

COMMERCIALIZATION OF THE BULBOSUM METHOD IN BARLEY (<u>HORDEUM</u> <u>VULGARE</u> L.)

L.P. SHUGAR

W.G. Thompson & Sons Limited
Blenheim, Ontario, Canada
NOP 1AO

M. Maluszynski (ed.), Current Options for Cereal Improvement, 161–170.
© *1989 by Kluwer Academic Publishers.*

162

ABSTRACT

This presentation includes a brief introduction of W.G. Thompson & Sons Limited, it's service, brokerage, seed sales and research efforts in Canada and abroad; some cereal research efforts were described. The paper describes the history of the doubled haploid technique in one commercial venture managed by 3 firms in concession over a 13 year period. The number of lines licensed using the technique and number of lines tested in the license trials are outlined as well as the number of lines yield tested since 1983 and projected to 1988. Breeding objectives and effective changes to the breeding system and techniques used are also presented. Technique efficiencies are presented in tabular form with some efficiency percentages related to modifications by W.G. Thompsons.

Introduction

W.G. Thompson & Sons Limited is based in southwestern Ontario, Canada. The firm is privately owned by the Thompson families. From a humble beginning back in 1924 with one white bean elevator handling facility, it now owns and operates ten elevators throughout Ontario (250 employees) handling white beans and soybeans, maize, rapeseed, winter wheat (soft white, hard red), spring wheat, winter and spring barley, oats and six and two-rowed barleys. In addition, the firm also supplies seed of the aforementioned crops through it's elevators and seed division sales network, Hyland Seeds. Thompsons also supplies animal care products, bird food, popcorn, agricultural chemicals, fertilizer and consultant services. In addition, through our brokerage section and sales specialists, large volumes of crop species are shipped all over the world.

In 1976, W.G. Thompson began a small maize breeding programme, added a white and soybean varietal development programme in 1977 and research programme in cereals in 1978. The research programmes are within the Hyland Seeds Division and proprietory varieties in maize and bean are already finding markets in Canada, the USA and Europe.

Our spring cereal breeding programme since 1978 has focused on six-rowed barley, some two-rowed, and a very little of oats. Originally all our effort in winter cereals was on soft-white wheat types for the pastry trade, but since 1983 we have stepped up our efforts to produce high yielding, disease resistant red winters as well. We utilized pedigree, S.S.D. and bulk-mass selection techniques in all cereal species until December 1984 when we purchased the elevator, production facilities and research programme from Ciba-Geigy Ltd. at Ailsa Craig and Nairn, Ontario, respectively. Table I shows the hectarage in research plots before and after the purchase.

In the spring of 1985 we began a study and crossing programme in hybrid wheat utilizing a gametocide in co-operation with Shell Inc., USA.

A short history of the use of doubled haploid technique in one commercial venture.

In 1974, John A. Stewart, a seedsman and elevator operator in Ailsa Craig, Ontario, started a research programme in small grains, corn and beans with help from the Crop Science Department, University of Guelph, in cereals. The doubled haploid technique was utilized in barley. Keh-Ming Ho and his wife, Louisa, managed the barley and winter wheat varietal development programs respectively. In 1976, Ciba-Geigy Seeds Limited purchased the production, elevator and cereal research facilities. In 1979, Mingo, a high yielding, doubled haploid line, with somewhat weak straw, was licensed for sale in Canada. By 1982, Keh-Ming and Louisa moved to Ottawa to continue their breeding careers and Dr. Ken Campbell assumed the role of manager of the cereal programme.

In 1983, Rodeo two-rowed barley was licensed for sale, with a strong straw and plump seed DH cultivar.

We have recently decided to continue utilizing the DH technique in barley as some of our research goals are becoming realized. The Canadian government had provided incentive funds from 1974 to 1985 for this project, however since then we have been running this rather expensive programme, plus the other cereal programmes, without grant aid.

Number of lines licensed using the technique compared with the number of lines entered into the licensing trials.

Two varieties have been released in the twelve year history of this doubled haploid barley varietal development programme. To date, 29 six-rowed and 53 two-rowed barley cultivars have been tested in their respective licensing trials. One additional six-rowed line is currently in it's second year of testing in western Canadian licensing trials.

For 1986, there are 5 two-rowed and 4 six-rowed barley entries in the Eastern Canadian Licensing Trials. Competitive genotypes are certainly coming out of the programme. Our new material seems to be shorter and stronger strawed, shows variability for maturity and important disease resistance and many, at least in our trials, have good yield potential.

Number of lines yield-tested since 1983 and success measured against the best check.

W.G. Thompson & Sons have been screening Weibull's two-rowed barley lines in Canada since 1978 and in parts of the United States since 1982. We have slowly increased the number of lines tested throughout the year and are planning to expand the area of testing for 1987. We are also screening the two-rowed lines from New Zealand, England, USA, Finland and Yugoslavia. Because many of these cultivars perform well here, we will be diminishing our two-rowed breeding programme somewhat and concentrating more on six-rowed barley production.

Until 1984 no more than 22 lines of the two-rowed and the same in six-rowed barley surpassed the checks Rodeo, Birka and Leger. In 1985, 34 two-rows and only 19 six rows and in 1986, 62 two-rows and 119 six-rows outyielded their respective checks. Table II shows the number of lines tested for yield since 1983 and projected to 1988.

Breeding objectives and means to get there

Three breeders have now been running this programme in twelve years. These techniques hastens pure line development. Moving population means maintaining higher variability for desired traits and keeping long-term goals in focus requires a continuity in approach. K. Ming made a few more six-rowed over two-rowed crosses, Ken favoured (70%) two-rowed and I, with my realities, six-rowed cultivars (75%).

In 1987, a preponderance of two-rowed material will reach yield trial level and again in 1988 (Table II). Generally, six-rowed types are lower than two-rowed barley lines in the percentage of haploid embryos rescued per 100 florets emasculated. Recently, researchers found the same in callus generation experiments (Kott et al., 1987). These results are reflected in the efficiency Table.

Other objectives:
1. High yield potential and stability.
    Merits of using the doubled-haploid technique and outline of techniques have been reviewed by Kasha and Reinbergs (1975, 1981) and Kasha and Seguin-Swartz (1983).

2. Good disease resistance.
    Some disease organisms of importance are <u>Pyrenophora teres</u>, <u>Cochliolobus</u> <u>sativus</u>, <u>Erysiphe</u> <u>graminis</u>, <u>Puccinia</u> <u>hordei</u>, <u>Rhynchossporium</u> <u>secalis</u> and BYDV. Strong, single disease resistance genes seem to exist.

3. Strong, bright straw, good thresh-ability

4. Plump, clean kernels.

5. Short term objectives:
    (a) to isolate the best lines and introduce for entry into licensing trials.
    (b) to inspect the present laboratory procedures for efficiency and reliability The technique has been observed to be very sensitive to slightly altered environments (Pickering, 1982).

6. Long term objectives:
    Parents should be selected very carefully.
    (a) Specific combinations can move to F_2 before gamete selection and more 'general' crosses can move into the laboratory in F_1.
    (b) New material moved in either as the 50% or 25% contributor.
    (c) Lines should be compatible with the doubled haploid system. This can be a very difficult point to achieve, therefore improvement of techniques may be required.
    (d) Lines should be tested in as wide an area as possible.
    (e) The short term objectives should be checked regularly.

<u>Changes made to the doubled haploid techniques since 1984</u>
(changes related to Fig. 1)

1. All crossing was performed inside during the winter months rather than in summer when field work and observation is so important. All culms were left intact.

2. The growth rooms were revented, thus providing better air flow and temperature control, resulting in healthier plants. By using florescent tubes with similar light intensity and spectrum, a savings of 50% was achieved.

3. The detaching of F_1 culms after bulbosum pollination was discontinued. This resulted in healthier seeds and haploid plants, less contamination of embryos, a higher percentage of survival and much less effort in handling.

4. Three days after pollination F_1 plants were moved into a special well ventilated nursery room with extra lighting. The pollinating bags were removed.

5. All new material was screened in the nursery before entering yield trials.

6. There are no changes in culturing or doubling techniques yet. As a result of numbers 2-5 above, there were more haploid plants and doubled haploid lines per haploid seedlings treated by colchicine.

    The efficiencies of doubled haploid production from 1980-81 through the partial 1986-87 season is presented in Table III.

Brief Overview

1. The DH technique works and fits in well with other breeding activities.

2. Special attention should be given to:
    (a) the choice of parents and the number of doses with each mate
    (b) the number of random gametes per cross (are there enough?)
    (c) care and respect of system by technicians
    (d) the actual field work, which is of great importance.

3. Consider the "interaction" cultivar-techniques according to breeding objectives. Six-rows are generally worse than two-rows in efficiency of production.

4. An expensive, labour intensive operation demands as wide a market as possible.

References

FALK, D.E., GUERRERO, P.L., (1987) Breeding malting barley using haploid techniques. Plant Breeding Symp. Lincoln, N.Z. Feb. 18-23 (in press).

JENSEN, C.J. (1975) Barley monoploids and doubled monoploids. Techniques and experiences. In: Barley Genetics III, 316-345.

KASHA, K.J., REINBERGS, R., (1975) Haploidy and polyploidy and it's application in breeding techniques. Genetica Yugoslavia, 307-315.

KASHA, K.J., REINBERGS, E., (1981) Recent developments in the production and utilization of haploids in barley. In: Barley Genetics IV, 655-665.

KASHA, K.J., SEGUIN-SWARTZ, G., (1983) Haploidy in crop improvement. In: Cytogenetics of Crop Plants, 19-68.

KOTT, L.S., KOTT, E., HOWARTH, M., KASHA, K.J., (1987) A comparative study of initiation and development of embryogenic callus from haploid embryos of several barley cultivars: I. Development state of embryo explants and callus potential. Canadian Journal of Botany, 65 (in press).

PICKERING, R.A. (1982) The effect of pollination bag type on seed quality and size in Hordeum inter and intraspecific hybridization. Euphytica, 31: 439-449.

Table I - <u>Hectares in Cereal Research Plots</u>:

Species	Before 1985	1985 and 1986
Winter Wheats		
Nursery	0.5	2.0*
Yield Trials	2.0	5.0
Oats		
Nursery	0.1	0.2
Yield Trials	0.3	0.5
Barley		
Nursery	0.5	2.0
Yield Trials	3.5	7.5**

* 2 nurseries   ** increases included

Table II - <u>Number of Lines Tested</u>

	1983	1984	1985	1986	Projected 1987	1988
2-Rowed						
Preliminary & Obs.	129	306	288 (72)*	226 (84)*	300 (100)*	50-100 (100-150)*
Advanced	90	80	77 (47)*	90 (30)*	44 (22)*	50 (30)*
6-Rowed						
Preliminary & Obs.	140	211	256	231	150	400
Advanced	60	105	80	86	80	100

* Weibullsholm 2-rowed Barleys

Table III - Doubled-Haploid Production Efficiency

[7] 1986 %	Operation	% at Thompsons						
		1980 -1981	1981 -1982	1982 -1983	1983 -1984	1984 -1985	1985 -1986	1986 -1987
23	Embryos cultured per flower poll.	16	07 ++	11	29	20	17	24 +
56	Haploid plants per embryo cultured	15	19	15	09 ++	14	18	35 +
83	Chromosome doubling per haploid plant	28	57	65	50	75 +	73 +	?
--	Hybrid % of treated haploids	09	10	11	27 **	09	09	low
11	Doubled-haploids per flower pollinated	0.7	0.8	1.1	1.3	2.0!	2.0!!	?
--	No. of flowers pollinated	241,368	123,246	247,632	198,216	194,568	229,512	?
		65%*	57%	61%	68%	30%	90%	50%

some 2-row
carryover

* % of six-rowed barleys in crossing program.

+ changes for the better

++resulted from contamination

** use of Swedish H.bulbosum

!-!! actual increase in efficiency x 5

170

Fig. 1 - Doubled-Haploid System*

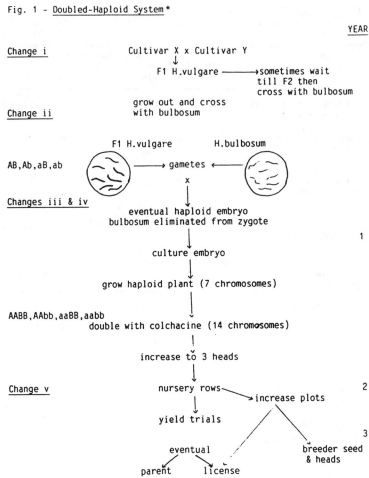

Change i   Cultivar X x Cultivar Y
                   ↓
             F1 H.vulgare ——→sometimes wait
                          till F2 then
                          cross with bulbosum

               grow out and cross
Change ii   with bulbosum

             F1 H.vulgare   H.bulbosum

AB,Ab,aB,ab   ——→ gametes ←——

                     x

Changes iii & iv

          eventual haploid embryo
         bulbosum eliminated from zygote   1

             culture embryo

      grow haploid plant (7 chromosomes)

AABB,AAbb,aaBB,aabb
        double with colchacine (14 chromosomes)

          increase to 3 heads

Change v   nursery rows   2
                       →increase plots

            yield trials

                            3

            eventual   breeder seed
                            & heads
       parent   license

*Modified figure from Jenson, C.J. 1975 [6]

INDUCED MUTAGENESIS TO FACILITATE HETEROSIS AND HYBRID SEED
DUCTION IN BARLEY

S.E. ULLRICH[1], M. MALUSZYNSKI[2,3], A. FUGLEWICZ[2]
AND A. AYDIN[1]

[1]Department of Agronomy and Soils
Washington State University, Pullman
Washington, USA

[2]Department of Genetics
Silesian University
Katowice, Poland

[3]Joint FAO/IAEA Division
Plant Breeding and Genetics Section
V i e n n a ,   A u s t r i a

M. Maluszynski (ed.), Current Options for Cereal Improvement, 171–183.
© 1989 by Kluwer Academic Publishers.

ABSTRACT

Hybrid barley (<u>Hordeum</u> <u>vulgare</u> L.) was produced on a limited basis in the 1970's in the western U.S.A., but ceased because of seed production problems and competition from newly released high yielding lodging resistant cultivars adapted to irrigated conditions. Reports of heterosis for yield in barley have been generally sufficient to justify hybrid cultivar development but variable over time and environments around the world, which indicates the choice of parents is critical. Several hybrid seed production systems have been investigated, and although none have been completely satisfactory, chemical hybridizing agents perhaps offer the greatest potential. Induced mutagenesis can play a role in hybrid barley development to enhance heterosis, female parent and hybrid seed production. To control height and lodging, semi-dwarf, stiff-strawed parents for barley hybrids will be imperative. Heterosis may or may not be expressed for plant height in F_1's from semi-dwarf mutant/mutant crosses. In studies with mutants from varied backgrounds several mutant/mutant crosses produced F_1's no taller than the taller mutant or shorter than either of the original progenitor cultivars.

---

* Scientific Paper No. 7699 . College of Agriculture and Home Economics Research Center, Washington State University, Project No. 1006. Research carried out in association with the IAEA under Research Agreeement No. 4468/CF.

1. INTRODUCTION

This presentation will include a short review of the information known about heterosis and hybrid cultivar development of barley (Hordeum vulgare L.) and a consideration of the use of induced mutagenesis to effect and enhance heterosis and hybrid seed production.

Barley production via hybrid cultivars has several economic prerequisites including a significant level of heterosis, an efficient and effective system of female parent seed production, and sufficient cross-pollination and therefore, hybrid seed production.

1.1. Heterosis. Heterosis for many characteristics of barley (e.g. growth, quality) may be important in hybrid cultivar development and use, but heterosis for grain yield will be the most important. Heterosis is defined here in relationship to the high parent. Since the first report of yield heterosis (27%) in barley (Immer, 1941), a wide range has been reported from below or near 0 (Lehman, 1981; Hagberg, 1953; Matchett and Cantu, 1977) to over 100% (Fejer and Fedak, 1976). Many European and North American researchers have reported yield heterosis levels between about 20 and 50% (see reviews by Scholz and Kunzel, 1981, 1986; Ramage, 1983). Overall it appears that these are acceptable levels of heterosis for yield in barley, although great variability exists. The choice of parents is critical and specific combining ability as well as general combining ability should be considered.

1.2. Hybrid Cultivar Development. The first and only commercial hybrid barley cultivars were grown on only 12,000-20,000 ha/year primarily under irrigation from 1970-1978 in the Southwest U.S.A. with a yield heterosis of 15-20% (Ramage, 1983). Major problems

were lodging and ergot (<u>Claviceps</u> <u>pupurea</u>). Hybrid production
ceased after the first short, stiff-strawed irrigated barley
cultivars such as 'Kombar' were released in the Southwest, U.S.A.
These new cultivars gave a yield increase similar to the hybrids,
when compared to the older taller cultivars.

Hybrid barley development has primarily centered on three
systems for female parent and hybrid seed production. Among the
first systems developed was one that uses balanced tertiary
trisomics (BTT) with coupled recessive genetic male sterile and
seedling lethal genes (Ramage, 1976). Other systems using genetic
male sterile and pre-flowering or pre-sowing selection genes (DDT
resistance, haploviables, blue aleurone, shrunken endosperm) have
been proposed (Ramage, 1983).

A second system to receive major attention uses cytoplasmic
male sterility and fertility restoration genes. Such systems have
been derived from crosses with <u>Hordeum</u> <u>jubatum</u> (Schooler and
Foster, 1968) and <u>Hordeum</u> <u>spontaneum</u> (Ahokas, 1978).

The third and most recently considered system uses male
gametocides or chemical hybridizing agents (CHA). Chemical
companies have expended considerable research and development
efforts in this direction. Whereas CHA's have shown great promise
and already limited use in wheat, none have been found to be
effective enough in barley to date, primarily because of dif-
ferential genotype response (Scholz and Kunzel, 1986).

Although considerable public and private research effort has
been expended over the last 20-25 years, sustained commercial
hybrid barley production is yet to be realized. It is probably a
matter of time, however. The use of F_1 hybrid barley cultivars
initially will be limited to high-input, high-yield areas

primarily, because of seed costs. However, relatively high yielding F_2 populations with less expensive seed costs could extend production into more marginal growing areas. Somewhat pessimistically, one could envision F_1 hybrid production being restricted to feed barley types as the complex of traits necessary for good malting and brewing quality could be difficult to achieve with heterozygous and heterogeneous F_2 seed, as was the case in the few studies conducted (Suneson and Dickson, 1966; Rasmusson et al., 1966; Foster and Schooler, 1970).

2. USE OF INDUCED MUTAGENESIS IN HYBRID BARLEY DEVELOPMENT

In spite of a brief period of commercial hybrid production, there are many refinements to F_1 seed production that would be enhancements or are prerequisite to sustained hybrid barley production. Induced mutagenesis could be an important tool to effect some of the changes.

2.1. Enhanced Heterosis. The expression of heterosis has been demonstrated in barley but not consistently. Selection of induced mutations to enhance heterosis has potential for example for overdominant expression (Aa>Aa or aa) for disease resistance, semi-dwarfism/lodging resistance with a dominant semi-dwarf gene, and for yield genes per se if they can be identified. Heterosis also could be enhanced with additional variability for cytoplasmic, as well as, nuclear genes and through nuclear-cytoplasmic gene interaction. Implied in these enhancements is improvement in specific combining ability as well as general combining ability.

2.2. Enhanced Systems for Female Parent and Hybrid Seed Production. The sources of cytoplasmic male sterility - nuclear restorer genes are very limited in barley. Additional genes for this system are desirable and likely could be induced. Induced

mutagenesis also could be used to search for genes that would
contribute to a self incompatability system, such as exists in rye
to develop a new hybrid system. Resistance to ergot has been
illusive, but the search could be renewed through induced
mutagenesis. This could have a significant impact on when and
where hybrid seed is produced as ergot has been a major problem.
Additional genes for resistance to smut, especially loose smut,
also would be useful.

Another major problem which could be overcome with induced
mutants is lodging of F_1 hybrids. Lodging is still a problem under
high yield conditions and would likely be increased with F_1 hybrid
production. To control plant height and lodging, semi-dwarf
stiff-strawed parents for barley hybrids will be imperative. Many
semi-dwarf barley sources are currently available (Konzak et al.,
1984, Ullrich and Aydin, 1985; Maluszyŝki et al., 1986). However,
a good dominant or cytoplasmic semi-dwarf is not available. It
would be useful to induce dominant and cytoplasmic semi-dwarf
mutant genes in barley to act in F_1 hybrids. An F_1 hybrid that
results from one or both parents that have a recessive gene(s) for
semi-dwarfism may be tall and prone to lodging due to dominance or
heterosis for plant height. Heterosis may or may not be expressed
for plant height in F_1's from semi-dwarf mutant/mutant crosses or
mutant/normal backcrosses, Tables I-IV. It should be noted that in
several cases mutant/mutant crosses produced F_1's no taller than
the taller mutant and/or shorter than either of the original
progenitor cultivars. Whereas heterosis for plant height may not
be desirable, some enhancement for height may be desirable if both
parents are dwarf or semi-dwarf types. The highest yield in barley
may come from the tallest plants without lodging (Ramage, 1984).

Heterosis for other characters such as tillering and grain production traits is desirable and occurred from mutant/mutant crosses or mutant/normal backcrosses even within the same background genotype, Table I. Our observations of mutant/mutant F_1's in the field indicate considerable variation for heterosis expression, and in some cases the heterosis or lack thereof would be desirable in hybrid production.

Semi-dwarf types will be useful in hybrid cultivar development, but many of the mutants available carry deficiencies, for example, reduced culm diameter which could affect lodging resistance (Ullrich and Aydin, 1985). Increased culm diameter and stiff straw should be selected along with decreased plant height. Alternatively particularly good mutants otherwise could be treated with a mutagen for the induction of increased culm diameter. Another deleterious trait that often occurs in semi-dwarf mutants is reduced seed size (Ullrich and Aydin, 1985). In general, the size of a plant part is proportional to the size of the meristem from which it arises, and in cereals seed size tends to be proportional to lemma and palea size. Therefore, by selecting within mutagenized semi-dwarf populations, one could directly select for large seed size or indirectly by selecting for increased culm diameter or lemma and palea size.

2.3. Enhanced Cross Pollination. Of paramount importance in hybrid cultivar development is sufficient seed set on the female (male sterile) parent line. Barley is highly self-pollinating. Therefore, traits favoring cross-pollination are prerequisite. Some of these traits are:

Anthesis after head emergence

Anther extrusion

Large anthers, more pollen/anther

Long duration anthesis-stigma receptivity

Delayed pollen dehiscence until after anther extrusion

Stigma extrusion - large stigmas

- small palea

Reduced awn length

Some of these traits (awn length, anther, stigmas extrusion) are more easily selected for than others (anthesis after spike emergence, long duration anthesis), therefore, screening techniques would have to be developed for the latter. Several of these traits were discussed in the 3rd FAO/IAEA RCM for semi-dwarfism in cereals in Mexico, 1984.

The development and use of male sterile facilitated recurrent selection (MSFRS) has both intentionally and inadvertently enhanced both cross pollination and heterosis potential in barley. The CCXXX population was released as a source from which cross pollinating lines could be selected under different environments (Ramage et al., 1976a). Male sterile lines derived from this population in North America and Europe have set seed up to 80-90+% (Ramage, 1983). The CCXXXII population was primarily developed as a semi-dwarf source for high yield conditions and hybrid cultivar parent development (Ramage et al., 1976b). Our observations under Washington and Arizona conditions indicate a high average percentage of seed set on male sterile plants in CCXXXII. Induced mutants could contribute to the cross-pollinating ability and heterosis in barley as is, and in conjuction with MSFRS techniques. Improved induced mutants could be incorporated into existing or new MSFRS populations.

TABLE I. The effect of heterosis measured as percent increase
of F_1 plants compared to parent varieties from crosses
involving semi-dwarf mutants of barley, Poland.

Cross	Plant height	Tillering	Grains per plant	Grain weight per plant	1000 Grain weight
Backcrosses					
A x 282	12.9*	61.0*	92.9*	113.7*	10.6
A x 233	16.1*	66.0*	104.5*	83.6*	- 8.9
A x 243	11.8	76.1*	80.6*	94.1*	7.4
T x 188	4.9	42.2	60.7*	143.5*	33.8*
D x 125	-1.5	96.8*	78.3*	55.4	-12.9
Diva Mutant Intercrosses					
243 x 233	13.7	58.5*	91.7*	60.8*	- 8.9
233 x 282	12.6	66.7*	178.7*	217.6*	13.8
233 x 239	11.6	73.6*	127.3*	133.3*	2.6
282 x 239	12.9	25.8	130.4*	152.9*	9.8
282 x 243	5.9	71.1*	7.3	11.8	4.0
243 x 239	7.8	-8.2	45.2*	35.3	- 6.9
275 x 233	11.7	72.9*	74.5*	56.9*	-10.1

A - parent variety 'Aramir', T - parent variety 'Trumph',
D - parent variety 'Diva'.
* - statistically significant

TABLE II. Culm length comparisons of parent and F_1 populations from semidwarf barley crosses, Washington.

Populations	Female (cm)		Male		F_1		t values comparing:			
	Mean	Range	Mean	Range	Mean	Range	F_1 & Female	F_1 & Male	F_1 & N1	F_1 & N2
Advance x A1	57.6±3.3	51-62	47.7±2.3	44-51	53.8±3.3	50-57	2.08	4.32**	---	---
A1 X Advance	47.7±2.3	44-51	57.6±3.3	51-62	56.7±3.3	62-61	7.50**	0.59	---	---
Advance x A2	52.8±4.4	44-59	43.1±3.0	37-50	49.5±5.0	41-55	1.94	4.33**	---	---
Morex x MO2	63.9±3.4	56-71	50.9±2.9	46-56	62.0±4.3	52-68	1.76	10.96**	---	---
Morex x MO4	61.6±4.4	52-68	37.0±3.2	30-43	59.5±3.6	52-66	1.91	24.50**	---	---
AMT531 x MO1	65.0±7.1	51-78	48.9±3.5	41-56	74.5±2.1	73-76	1.85	10.25**	1.39	1.39
Manker x MA1	61.0±4.3	51-67	53.9±3.1	48-61	61.0±3.8	54-67	0.30	6.97**	---	---
Larker x L2	63.2±3.9	56-70	49.7±3.8	42-57	61.1±3.0	58-66	1.22	7.45**	---	---
Larker x L3	61.5±4.6	51-70	50.8±3.3	42-57	60.0±2.8	58-62	0.45	3.87**	---	---
Larker x L1	63.8±3.8	55-69	56.8±5.4	43-64	63.4±5.9	52-71	0.24	3.13**	---	---
WA9037-75 x WA371	65.4±4.3	56-72	55.4±2.4	51-61	63.4±3.8	60-73	1.43	8.35**	---	---
WA371 x MO6	58.4±3.7	51-66	61.6±3.7	55-67	68.9±7.8	53-84	5.66**	3.75**	1.71	3.02**
WA9044-75 x WA441	60.5±4.6	53-70	46.3±2.3	43-54	59.4±2.8	55-64	0.75	15.17**	---	---
Norbert x N1	63.5±3.7	58-72	53.1±3.7	46-60	61.8±4.5	57-71	1.12	5.38**	---	---
MO6 x N1	60.8±3.2	55-67	51.7±3.7	46-62	64.0±4.5	58-71	2.57*	9.41**	1.54	0.35
N1 x MO6	54.9±4.5	46-62	64.3±3.7	56-70	74.6±3.8	67-80	13.70**	8.68**	9.21**	9.73**
Harrington x H1	55.2±2.7	50-60	47.8±3.6	42-54	54.3±2.6	51-58	1.04	5.79**	---	---
WA10698-76 x WA981	46.3±3.6	40-53	42.3±2.4	38-46	44.1±3.6	40-50	1.58	1.56	---	---

*,** Differences significant at the 0.05 and 0.01 level of probability, respectively.
N1, N2 normal isotypes of the female and male mutant parent, respectively, for each cross.

TABLE III. Culm length comparisons of parents and F₁ populations from crosses among semidwarf barley mutants, Washington.

Populations	Female Mean	Female Range	Male Mean	Male Range	F₁ Mean	F₁ Range	t values comparing: F₁ & Female	F₁ & Male	F₁ & N1	F₁ & N2
A1 x L2	41.6±2.4	37-47	47.3±2.8	40-52	48.9±3.1	43-54	7.67**	1.56	2.87**	8.88**
A2 x L3	44.0±4.1	36-52	53.6±4.2	46-59	52.4±4.1	43-58	6.58**	0.79	0.27	6.10**
L2 x A2	51.0±3.6	42-58	39.5±2.7	33-44	49.0±1.9	46-52	1.76	10.63**	8.57**	2.74**
L4 x A2	54.5±3.7	47-60	42.4±3.5	36-47	53.7±3.6	48-59	0.70	9.54**	5.18**	0.55
M02 x M04	56.5±4.8	46-64	42.0±4.0	34-49	60.9±5.6	52-69	2.23*	10.84**	1.99	1.99
M04 x M02	42.0±4.7	31-48	54.5±5.3	45-62	62.4±5.6	50-73	11.80**	4.38**	1.18	1.18
M03 x M02	53.5±4.6	45-62	50.4±4.3	43-60	61.1±3.8	54-67	5.88**	9.18**	0.39	0.39
L3 x L2	53.7±3.1	46-60	55.0±3.2	48-61	54.6±3.3	48-60	0.92	0.35	4.89**	4.89**
L3 x L4	53.7±3.1	46-60	51.6±2.9	46-57	52.5±2.6	48-58	1.16	0.93	6.17**	6.17**
L2 x MA1	49.0±5.5	36-57	50.8±4.3	42-60	52.0±4.2	43-59	1.85	0.84	6.34**	6.60**
L4 x MA1	51.0±4.8	41-59	51.2±4.6	40-59	52.8±2.6	49-58	1.39	1.37	4.81**	4.83**
L1 x MA1	55.4±4.4	49-65	47.8±4.5	39-59	59.2±2.4	54-65	3.82*	10.94**	2.08*	1.78
M04 x M05	41.9±3.3	33-47	60.8±4.1	51-69	61.0±4.4	56-66	10.37**	0.11	1.59	1.59
M06 x M03	62.7±4.7	54-70	54.3±4.8	43-61	63.0±4.2	56-68	0.19	6.23**	0.92	0.92
M06 x M04	63.2±4.6	51-70	36.5±3.5	29-40	61.0±5.2	53-74	1.60	17.18**	0.39	0.39
M04 x M06	40.3±4.2	29-47	60.5±4.6	47-68	68.0±0.0	68-68	6.48**	1.58	1.41	1.41
N1 x M01	56.4±3.2	49-62	46.6±4.4	38-54	69.5±3.0	64-74	11.92**	16.22**	4.89**	5.50**
WA371 x M02	57.8±4.1	46-65	55.8±2.7	47-61	67.0±3.1	60-72	7.85**	13.14**	1.32	3.18**
WA371 x M03	59.4±3.2	54-66	52.9±3.5	48-59	69.3±2.6	62-74	10.10**	17.22**	3.41**	5.89**
L1 x WA371	55.0±3.7	45-60	56.5±3.1	50-62	66.4±3.7	62-74	8.15**	8.37**	1.84	0.64
WA371 x WA441	55.1±3.1	47-61	48.2±2.3	44-52	56.9±3.2	52-63	1.94	10.78**	6.85**	2.70**
WA981 x WA371	45.2±3.6	39-53	55.4±3.2	48-61	53.5±2.7	49-57	7.32**	1.70	6.09**	8.67**
H1 x MA1	51.2±3.0	45-58	53.6±3.5	46-60	74.0±5.6	62-81	16.32**	14.17**	14.88**	7.73**

*, ** Differences significant at the 0.05 and 0.01 level of probability, respectively.
N1, N2 normal isotypes of the female and male mutant parent, respectively, for each cross.

TABLE IV. Culm length comparisons of parents and F_1 populations from semidwarf barley crosses, Washington.

Populations	Female (cm)		Male (cm)		F_1 (cm)		t values comparing:			
	Mean	Range	Mean	Range	Mean	Range	F_1 & Female	F_1 & Male	F_1 & N1	F_1 & N2
L3 x MO5	50.6±4.3	43-57	58.4±5.7	49-68	61.4±3.9	54-68	8.40**	1.94	0.78	2.41*
MO5 x A1	56.7±4.1	49-64	41.5±2.5	33-45	54.9±4.2	47-62	1.16	11.56**	4.01**	1.21
MO2 x MA1	57.0±4.8	48-66	54.3±5.3	43-65	68.6±5.7	59-84	7.62**	8.63**	3.62**	4.93**
MO2 x L1	55.2±6.0	45-64	60.8±3.2	55-66	64.7±5.8	57-76	4.93**	2.71**	0.56	1.82
MO4 x L2	36.7±5.3	25-46	49.8±2.7	46-56	57.8±2.9	53-61	10.60**	7.21**	2.23*	2.10*
MO6 x L1	62.4±4.5	53-70	55.3±3.6	47-62	61.0±3.6	53-65	1.08	5.20**	0.45	0.39
MO3 x MA1	54.4±3.7	48-60	51.0±3.7	47-60	63.5±4.0	58-72	7.56**	10.28**	1.40	1.88
MA1 x MO6	44.8±4.3	36-51	58.9±5.4	47-66	57.5±4.1	50-63	9.32**	0.91	2.82**	3.13**
MA1 x MO4	45.4±4.7	35-53	35.8±4.5	26-44	57.9±4.5	50-63	6.16**	11.10**	1.71	1.94
L1 x MO3	52.0±4.5		54.0±3.2	47-59	62.0±5.7	52-71	5.85**	5.47**	1.11	0.22
MO4 x L1	39.7±3.4	34-46	57.6±5.0	50-70	58.9±5.6	50-68	13.37**	0.74	1.52	1.40
N1 x MO2	57.9±6.0	43-65	56.7±4.4	48-65	74.1±4.9	66-84	8.70**	12.27**	7.88**	8.31**
Manker x MO2	65.9±7.2	53-76	55.8±4.7	47-63	67.4±5.3	57-78	0.61	6.54**	----	2.64**
WA441 x N1	48.7±2.7	44-53	51.4±4.1	43-59	57.2±4.5	50-65	7.35**	4.20**	2.10*	4.90**
WA441 x MO6	48.4±3.9	41-59	61.4±3.3	53-67	65.7±3.8	60-74	13.61**	3.63**	3.62**	1.66
H1 x WA371	49.6±2.9	44-54	56.5±3.2	50-62	60.4±3.3	54-66	11.97**	3.81**	7.92**	4.30**
Harrington x WA981	53.1±3.2	45-60	45.5±3.7	38-52	56.6±3.9	49-63	3.40**	9.22**	9.15**	----

*, ** Differences significant at the 0.05 and 0.01 level of probability, respectively.
N1, N2 normal isotypes of the female and male mutant parent, respectively, for each cross.

LITERATURE CITED

(1) AHOKAS, H., Cytoplasmic male sterility in barley. I.
     Anther and pollen characteristics of msm1, restored and
     partially restored msm1 genotypes. Z Pflanzenzuechtg
     81 (1978) 327.
(2) FEJER, S.O., FEDAK, G. Heterosis and combining ability in a
     diallel cross of six-rowed spring barley selections. Barley
     Genet III (1976) 797.
(3) FOSTER, A.E., SCHOOLER, A.B., Cytoplasmic male sterility in
     barley. Barley Genet II (1970) 316.
(4) HAGBERG, A., Heterosis in barley. Hereditas 39 (1953) 325.
(5) IMMER, F.R., Relation between yielding ability and
     homozygosis in barley crosses. J Am Soc Agron 33 (1941) 200.
(6) KONZAK, C.F., KLEINHOF, A., ULLRICH, S.E., Induced mutations
     in seed-propagated crops. Plant Breeding Reviews (Janick J,
     ed.) 2 (1984) 13.
(7) LEHMANN, L., Where is hybrid barley? Barley Genet IV (1981)
     772.
(8) MALUSZYNSKI, M., MICKE, A., SIGURBJORNSSON, B., SZAREJKO, I.,
     FUGLEWICZ, A., The use of mutants for cross-breeding and
     for hybrid barley. Barley Genet V In press (1986).
(9) MATCHETT, R.W., CANTU, O.P., Hybrid barley and an illusive 8
     year chase. Barley Newsletter 20 (1977) 130.
(10) RAMAGE, R.T., Hybrid barley. Barley Genet III (1976) 761.
(11) RAMAGE, R.T., Heterosis and hybrid seed production in
     barley. Chap 3. Heterosis (Frankel R, ed.). Monograph on
     Theonet Appl Genet 6 (1983) 71.
(12) RAMAGE, R.T., Status of hybrid barley. IAEA RCM Mexico
     (1984).
(13) RAMAGE, R.T., THOMPSON, R.K., ESLICK, R.F., WESENBERG, D.M.,
     WIEBE, G.A., CRADDOCK, J.C., Registration of barley
     composites XXX-A to -G. Crop Sci 16 (1976a) 314.
(14) RAMAGE, R.T., THOMPSON, R.K., ESLICK, R.F., Release of
     Composite Cross XXXII. Barley Newsletter 19 (1976b) 9.
(15) RASMUSSON, D.C., UPADHYAYA, B.R., GLASS, R.L., Malting
     quality in F_1 hybrids of barley. Crop Sci 6 (1966) 339.
(16) SCHOLZ, F., KUNZEL, G., Progress and problems with hybrid
     barley. Barley Genet IV (1981) 758.
(17) SCHOLZ, F., KUNZEL, G., Hybrid barley-problems and advance,
     especially in developing genetic systems. Barley Genet V In
     press (1986).
(18) SCHOOLER, A.B., FOSTER, A.E., Cytoplasmic male sterility
     found in barley. Barley Newsletter 11 (1968) 18.
(19) SUNESON, C.A., DICKSON, A.D., Hybrid barleys are coming.
     Master Brewers Assoc Am Tech Q 3 (1966) 185.
(20) ULLRICH, S.E., AYDIN, A., Mutation breeding for semi-
     dwarfism in barley. IAEA RCM, Rome. TECDOC In press (1985).

SOME PROSPECTS OF HYBRID TRITICALE BREEDING

WOLSKI, T., POJMAJ, M.S.

Poznan Plant Breeders
Warsaw, /Poland

M. Maluszynski (ed.), Current Options for Cereal Improvement, 185–192.
© *1989 by Kluwer Academic Publishers.*

ABSTRACT:
Results obtained by wheat and rye breeders as well as reports on hybrid vigour in triticale led us to question the levels of heterosis to be achieved in our triticale breeding material. Single crosses between triticale lines and varieties of different origin were examined in the mean of 1000 kernel weight. In some of them a marked heterosis in this trait was observed (16.4% over the parent mean).
Because of the lodging problem, interest was taken in the performance per se of short-stemed and lodging resistant strains. The data are presented. The incorporation of dwarfing genes into triticale might be valuable. The EM 1 dwarfing gene from rye seems to be of high yield. The mutation progamme for broadening variability in stem shortness and in combining ability was initiated.

INTRODUCTION

Triticale is becoming an important cereal crop in Poland. It is estimated that around 250,000 ha winter triticale have been planted in the fall of 1986 (as a substitute for rye) and 1 million are anticipated before 1990.

According to our observations F_1 triticale shows hybrid vigour, which is confirmed by several authors (Bell, 1985; Gill et al., 1979; Guzhov et al., 1982; Srivastava and Arunachalam, 1977). Studies on diallel crosses show that nearly all characters connected with yield display a specific combining ability (Carrillo et al., 1983; Gill et al., 1979; Kaltsikes and Lee, 1973; Rao and Joshi, 1979; Reddy, 1976). Another important factor is the larger pollen distribution of triticale as compared to wheat (D'Souza, 1970, Sapra & Hughes, 1975; Young and Larter, 1972).

Furthermore, different wheat and rye dwarfing genes, some of them dominant, are available in the triticale breeding material, which seems important to obtain lodging resistant hybrids.

In recent years the revived interest of wheat breeders in utilization of heterosis may be observed, owing to chemical gametocides. Thus, screening of large numbers of lines for combining ability became possible as well as commercial production of hybrid seed. All problems connected with the use of cytological male sterility are avoided.

Most interesting data on heterosis in wheat crosses were presented (Gale et al., 1986; Austin et al., 1986). On the grounds of these results hybrid vigour in grain yield, as high as 8.6%, may be expected. It was entirely attributed to heavier kernels. The use of Rht3 parents increased the number of grains per ear, which is in this case a dominant factor. However, the uniformity of the hybrids may be a problem, because of tall aneuploids which may segregate in proportions higher than standard for seed certification.

RESULTS

In 1986 in Choryn Plant Breeding Station the 1000 kernel weight of 81 F_1 single-cross hybrids was compared to that of their 31 parents (Table 1). Among them only one line, derived from a cross with the Ukrainian variety AD 206, has given a marked heterosis in 1000 - grain weight, both as seed bearing and male parent. This is in accordance with the results of Guzhov et al., (1982) who demonstrated the good combining ability of AD 206. In our case heterosis was as high as 16.4% (as compared to the parental means) as the average of 9 cross-combinations with the given line used as male parent and 3 crosses in which it was used as the seed-bearing parent.

An essential factor for obtaining successive hybrids is the lodging resistance and short straw of both or at least one parent. In the case of dominant or semi-dominant dwarfing genes the female parent should be short and the male parent may be quite tall. At least two sources of such a character are available in triticale. One is the well known Rht3 Tom Thumb wheat gene, which proved rather unsuccessful in conventional breeding. The second one is the EM 1 semi-dominant rye dwarfing gene, transferred more recently to triticale (Wolski and Tymieniecka, 1988). This material, at present in the F_1 generation, seems more promising. Numerous lines are characterized by plump and big grain and the plant architecture is fully acceptable.

Furthermore, a number of short stem lines were obtained in recombination breeding. They were tested in 1985 in Laski and Choryn Stations and in 1986 in Laski and Debina Stations, the latter being located on very rich soil. The results concerning some interesting lines are presented in Tables 2 and 3, in comparison to varieties Bolero in 1985 and Salvo in 1986. Two lines are short and very resistant to lodging, LT 3102_{83} tested in both years and CT 829_{83} - in 1985 only. The former is derived from a cross with CIMMYT material (274/320 - Bunny), the latter - with a semi-dwarf wheat, the shortness being presumably attributed to CIMMYT Norin 10 genes. According to observations both behave in most cross-combinations as semi-dominant, therefore both deserve attention from a practical and theoretical point of view. Another interesting line is LT 1439_{82} - not as short as the former ones but relatively lodging resistant and giving a very high yield in Debina.

We believe that the genetic background of the existing sources of short stem and lodging resistance in triticale should be better recognized, including GA response. The spectre of this germ-plasm should also be broadened, with the use of mutation techniques.

CONCLUSIONS:

1. In the light of the published data on hybrid vigour of triticale and the possibility of using gametocides, research in this field seems necessary.

2. It is possible to find triticale lines displaying hybrid vigour in 1000 kernel weight, although it cannot be regarded as a general phenomenon.

3. As triticale hybrids should be relatively short and resistant to lodging, parental lines of this type should be used.

4. There exists a number of sources of the short stem and semi-dwarf character in triticale. They need genetic identification.

5. Broadening of the spectre of short and semi-dwarf triticales by means of mutation breeding is required.

References

AUSTIN, R.B., FORD, M.A., MORGAN, C.L. (1986) Yield, yield components and straw yields of F_1 hybrid winter wheats and their parents. Plant Breeding Institute, Cambridge, Annual Report 1985: 109-110.

BELL, R.K. (1985) Heterosis for important characters in hexaploid triticale. Wheat Inf. Service 60: 10-14.

CARRILLO, J.M., MONTEAGUDO, A., SANCHEZ-MONGE, E., (1983) Inheritance of yield components and their relationship to plant height in hexaploid triticale. Z. Pflanzenzücht. 90: 153-165.

D'SOUZA, L., (1970) Untersuchungen über die Eignung des Weizen als Pollenspender bei der Fremdbefruchtung, verglichen mit Roggen, Triticale und Secalotricum. A. Pflanzenzücht. 63: 246-269.

GALE, M.D., HOODGENDORN, J., SALTER, A.M., (1986) The exploitation of Rht3, Tom Thumb dwarding gene, in F_1 hybrid production. Plant Breeding Institute, Cambridge, Annual Report 1985: 67-69.

GILL, K.S., BHARDWAJ, H.L., DHINDSA, G.S., (1979) Heterosis and combining ability in triticale. Cereal Res. Comm. 4: 355-362.

GUZHOV, Y.L., VELLANKI, R.K., MAKSIMOV, N.G., (1982) Prospects of breeding triticale for heterosis. Doklady Vsesoyuznoi Ordena Lenina i Ordena Trudovogo Krasnogo Snameni Akademii Sel'sko - khozyaistvennykh Nauk Imeni V.I. Lenina 9: 6-8 (in Russian).

KALTSIKES, P.J., LEE, J., (1973) The mode of inheritance of yield and characters associated with it in hexaploid triticale. Z. Pflanzenzücht; 69: 135-141.

RAO, V.R., JOSHI, M.G. (1979) A study on inheritance of yield components in hexaploid triticale. Z. Pflanzenzücht. 82: 230-236.

REDDY, L.V., (1976) Combining ability analysis of some quantitative characters in hexaploid triticale. Theor. Appl. Genet. 47: 227-230.

SAPRA, V.T., HUGHES, J.L., (1975) Pollen production in hexaploid triticale. Euphytica 24: 237-243.

SRIVASTAVA, P.S.L., ARUNACHALAM, V., (1977) Heterosis as a function of genetic divergence in triticale. Z. Pflanzenzücht. 78: 269-275.

WOLSKI, T., TYMIENIECKA, E., (1988) Effect of the rye
    mutant EM 2 dwarfing gene on octo- and hexaploid
    triticale (preliminary report).   In: Semi-dwarf cereal
    mutants and their use in cross breeding III, IAEA,
    TECDOC-455: 247-254.

YOUNG, K.C., LARTER, E.N. (1972) Pollen production and
    disseminating properties of triticale relative to
    wheat. Can. J. Plant Sci. 52: 569-574.

Table 1.   Hybrid effect   of 1000-kernel weight in some
           F1 triticale crosses / Choryń 1986 /   .

	Number of cross combinations	1000-k weight of the parent	Mean 1000-k weight of both parents	Difference mean %%
Average	81	53.7		+ 2.2
Highest	12	48.2	49.5	+ 16.4 ± 7.7
Lowest	10	64.7	57.5	− 6.3 ± 6.8

Table 2. Characteristic of two short triticale lines as compared with cv. Bolero in a trial at two sites in 1985. /4 reps, 5m^2 plots/

Variety or line	Grain yield t/ha			Plnt ht cm	Lodg*	Leaf dise ases*	Test wt kg/hl	1000 g.w. g	Fall no sec	Prot %
	Laski	Choryń	Mean							
Bolero	9.05	8.63	8.84	111	4.4	4.0	70.4	38	83	11.3
LT 310283	8.10	9.27	8.69	93	1	5.5	69.5	37	155	12.6
CT 82983	6.40	6.35	6.38	74	1	7.5	62.6	43	115	13.4
LSD	0.86	1.13	1.09							

* 9 – point scale / 1 – the best /

Table 3. Characteristic of two short and one lodging -
- resistant triticale lines as compared with
cv. Salvo / also resistant to lodging / in a
trial at two sites in 1986 /4 reps, $5m^2$ plots/.

Variety or line	Grain yield t/ha			Plnt ht cm	Lodg *	Test wt kg/hl	1000 g.w. g
	Laski	Dębina	Mean				
Salvo	6.08	9.80	7.94	116	2	63.2	46
LT 1439$_{82}$	7.41	10.74	9.08	115	1.7	69.0	45
LT 3102$_{83}$	6.43	8.45	7.44	95	1	64.8	42
LT 1225$_{84}$	7.44	8.88	8.16	104	2	65.6	39
LSD	0.60	0.30	0.50				

* 9 - point scale / 1 - the best /

INDUCTION OF SALT TOLERANCE IN HIGH-YIELDING RICE VARIETIES THROUGH MUTAGENESIS AND ANTHER CULTURE

F.J. ZAPATA and R.R. ALDEMITA

International Rice Research Institute
P.O. Box 933
Manila, The Philippines

M. Maluszynski (ed.), Current Options for Cereal Improvement, 193–202.
© *1989 by Kluwer Academic Publishers.*

Abstract

The practical application of androgenic haploid production to varietal improvement is hampered in rice by the low response of high-yielding _indica_ varieties to anther culture.   Recent advances using the combined techniques of radiation and in vitro culture show good possibilities to improve their androgenic ability.   It was demonstrated that minimal stress on the explant by irradiation may not induce irreversible genotypic changes but can stimulate callus induction and plant regeneration.   Because of the absence of green plant regeneration from panicles derived callus of rice variety Basmati 370, gamma radiation was applied to the seeds.   The results show that radiation can trigger green plant regeneration. The agronomic characters of the anther culture derived lines indicated that through irradiation and subsequent anther culture, useful variability may arise.

INTRODUCTION

The production of haploids has proven to be a very useful and efficient method of producing homozygous inbred plants, detection and recovery of recessive mutants and also in genetic analysis and selection within small population (Nitsch and Godard, 1979; Foroughi-Wehr et at., 1982). Anther culture which is one of the methods of obtaining haploid plant is the most widely used technique of haploid production. The technique involves primarily the induction of pollen grains from cultured anthers to give rise to androgenic haploids through a marked shift in the pattern of their normal development. Following the first mitosis which may or may not be normal, a haploid callus tissue or embryoid is formed instead of the normal male gametophyte. There is then a series of physical, physiological, nutritional and genetic factors controlling the development of haploids in culture.

However, the practical application of androgenic haploid production to varietal improvement is hampered by the low response of high-yielding indica varieties to anther culture. Techniques to improve the culturability of crops have been employed, like physical and chemical stresses, and the use of complex mixtures of organic origin. Another method through which improvement in plant regeneration can be obtained is through the application of stress in the form of ionizing radiation.

Recent advances using the combined techniques of radiation and tissue culture show good possibilities. Investigations on seed germination and response in culture of isolated hypocotyl segments of pigeon pea (Cajamus cajan) subjected to gamma radiation from 0-10 kr using ^{60}Co have shown stimulation of regenerative potential (Sharma Rao and Narayanaswamy, 1976).

Induced mutation is used as a complementary tool in plant breeding for crop improvement: its first advantage is that the basic genotype of the variety is usually only slightly altered while the improved character is expressed. Secondly, induced mutations can also offer a method of breaking tight linkages producing chromosomal translocations for gene transfer. In rice, the subspecies indica are known to be slightly amenable to in vitro culture. This could be due to the tight linkage between the non-culturability character and the agronomic characters inherent to the indica rices. By breaking the linkage through radiation it may be possible to express the character for culturability. Thirdly, polygenic changes for the improvement of quantitative characters like yield can also be induced. More important, additional variability and synergistic effect which may or may not be found in the natural population may also be obtained by subjecting F_1 seeds to mutagenic treatment, particularly in interracial crosses (Sigurbjörnsson, 1968).

OBJECTIVES:

1.   Improvement of culture ability of recalcitrant
      high-yielding varieties through radiation.

      Some modern rice varieties have many positive
characters such as high yielding capacity and good plant
type, but in many instances are susceptible to some pests
and diseases, and to some extent, to environmental
stresses.   This may be the reason why some of the modern
rice varieties do not replace the traditional varieties in
farmers' fields.   Improvement of indica varieties by
radiation followed by anther culture may be fruitful due to
possible induction of variabilities and the immediate
fixation of homozygosity.   However, most of the indicas are
recalcitrant and are thus not amenable to in vitro culture.
This may be due to the presence of some genetic linkages
between the culture ability character and the desirable
agronomic traits.

      Very often, application of ionizing radiation-like
gamma rays could stress the plant resulting in some
cytological and molecular changes such as chromosomal
aberrations, translocations and breaking of linkages.   It
may be possible that minimal stress on the explant by
irradiation may not induce irreversible genotypic changes
but only induce callus induction and plant regeneration.

2.   Induction of salt tolerance in high yielding rice
      varieties.

      One of the constraints to increasing rice production is
that many rice lands suffer from soil problems.   Among these
problems, soil salinity is the most predominant.   In South
and Southeast Asia alone, about 58.8 million hectares of
rice lands are cultivated with poor results because of salt
injury (IRRI, 1980).

      Some rice varieties with salt tolerance have been
identified but all are traditional cultivars with low
yielding ability.   Improving the yield of these varieties
has been attempted but due to several reasons, such as high
susceptibility to pests and diseases, existence of
undesirable genetic linkages between adaptability and
productivity, etc., progress is very limited.   As a
consequence, there is an increasing emphasis on improving
salt tolerance in modern high yielding varieties through
mutagenesis and tissue culture.

      The objective of this experiment is to induce salt
tolerance in modern rice varieties through the use of
radiation and tissue culture (in vitro cell selection and
anther culture).

3.   Field evaluation of anther culture derived lines.

      Irradiation could possibly increase variability which
can be fixed through anther culture.   Field evaluation of
the anther culture-derived lines would make it possible to
select and identify new variants that may have the same

agronomic characters and high yield as the parent plant but with other desirable characters like resistance to pests and diseases and environmental stresses. The objective therefore is to incorporate resistance to pests, diseases and environmental stresses into high-yielding varieties through the anther culture of irradiated modern cultivars and identify variants in terms of improved agronomic characteristics and tolerance to stresses.

METHODOLOGIES

1. Improvement of culture ability of recalcitrant high-yielding varieties.

    Seeds of high yielding IR varieties such as IR8, IR28, IR36, IR42, IR43, IR50, IR54, IR64 and other cultivars, Pokkali and Taipei 309 will be exposed to gamma rays from a ^{60}Co source at 0, 10, 20, 30 and 40 kr dosage at 1.87 kr per minute. Seeds will be sown and panicles collected before flowering. Anthers will be selected when the pollen grains are at middle uninucleate to early binucleate stage of development. They will be plated in a callus induction medium, incubated in the dark for 8 days at 8°C and later transferred to dim light conditions at 25°C. Calli formed will be transferred to a plant regeneration medium.

    Data on callus induction, percentage of calli producing plants, green and albino plant productions as affected by the irradiation treatments will be collected.

2. Induction of salt tolerance in high-yielding rice varieties through irradiation

(a) Some of the traditional rice varieties are salt tolerant. High yielding varieties will be crossed to these salt tolerant rice cultivars and the F_1 seeds will be irradiated at 0, 10, 20, 30 and 40 kr. Irradiated seeds will be anther cultured and induced to produce calli and green plants. Anther culture-derived lines will be evaluated for salt tolerance both in the glasshouse and in the field.

(b) Seeds of high yielding varieties will be irradiated to induce genetic alterations such as chromosomal aberrations and translocation. Seeds will be sterilized and cultured in a callus induction medium. Calli formed will be subjected to NaCl - stress by transferring them to callus induction medium containing various concentrations of NaCl or sea water. Surviving calli will be subjected to further stress in increasing concentrations of NaCl or sea water. After several passages of one month duration per passage, each surviving calli will be transferred to a regeneration medium for green plant regeneration. Regenerated plants will be screened both under glasshouse and field conditions.

3. Field evaluation of anther culture-derived lines

    Plants regenerated from anther culture will be seed increased. An observational yield trial will be conducted so as to evaluate and determine the extent of variabilities obtained. Anther culture-derived lines will be planted at one plant per hill in 5 rows of 5 m in length spaced at 20 cm between rows and hills. Two checks namely, the normal non-irradiated seeds and the irradiated seeds of each dose for each variety which were not subjected to anther culture, will be planted side by side with the anther culture derived lines.

Observations will be recorded every 7 to 10 days from tillering to flowering. Parameters to be used for evaluation will be yield components and agronomic characteristics.

Selected lines in the observational yield trial will be entered in the replicated yield trials for further evaluation. Lines will be grown and seedlings will be planted in 8 rows of 10 m long, with hills spaced at 20 x 20 cm, at one plant her hill. The most promising lines will be advanced to more rigid evaluation for varietal development.

ACHIEVEMENTS

The improvement of culture ability of the Basmati type rice through anther culture.

Basmati 370 is an aromatic rice with long, slender, excellent eating quality grains. The variety being tall and with weak and slender culm is prone to lodging. Thus, it will be important to obtain a variant which will have the grain quality of Basmati but is shorter and has a stiff culm.

Preliminary studies using panicles of non-irradiated Basmati 370 subjected to different callus induction media showed only albino plant regeneration (Table 1). The different media used affected the rate of callus and albino plant production. Medium E10 induced the highest callus production and albino plant regeneration. Because of the absence of green plant regeneration in non-irradiated seeds, ionizing radiation as gamma ray was applied to the Basmati seeds. The technique was done with the hope that irradiation stress will produce genetic alterations in the cell which would somehow improve callus induction and green plant regeneration. An added advantage of using anther culture will be the obtaining of stable plants which have been fixed in homozygous state.

MATERIALS AND METHODS

Seeds of Basmati 370 were subjected to irradiation dosages of 15, 20, and 25 kr with application dose rate of 1.2 kr/min from ^{60}Co gamma cell. Anthers from plants or irradiated seeds containing pollen at the middle uninucleate to early binucleate stages of pollen development were plated in G medium. Calli produced were transferred to a liquid Murashige and Skoog (MS) regeneration medium with 1 mg kinetin/l, 1 mg naphthaline acetic acid/l and 20 mg abscisic acid (ABA)/l for four weeks and then to ABA-free semi-solid MS regeneration medium (Torrizo and Zapata, 1986).

RESULTS

1. Anther culture

Both the percentage callus production and albino plant regeneration decreased with increasing dosage of gamma rays (Fig. 1). The lowest percentage values were obtained at 20 kr which eventually increased at 25 kr (Zapata and Aldemita, 1986). Green plant regeneration was observed only in 20 kr gamma ray. These results show that genetic alterations may have been induced by gamma irradiation which decreased callus production efficiency, and albinism, but at the same time triggered green plant regeneration. Although irradiation was directed randomly at the plant cells, genetic alterations manifested were favorable for green plant regeneration.

2. Field evaluation of regenerated plants.

Green plants regenerated from the foregoing experiment were evaluated in the observational yield trial in 1986 wet season. Table 2 shows the agronomic characters of the anther culture-derived lines from irradiated Basmati 370.

There is no significant difference on the days to flower among the anther culture lines and checks. In plant height, AC5821-2 and AC5821-3 were shorter than either the parent or the irradiated check. Decreasing the height of Basmati 370 is important to prevent lodging, however this decrease in height does not seem to be significant. Likewise, the average culm length and panicle length of 18 anther culture entries do not also differ significantly from both the parents and the irradiated check. The tiller number showed a wide range of variation, from 9 to 15 tillers per plant. However, the average value of 18 lines does not differ significantly from the parent. The percent fertility values vary among the anther culture lines. The average percent fertility of the 18 lines is much lower and significantly differ from the parent. The 100 grain weight varied in all the anther culture entries. The average however, does not differ significantly from the parent. It can be observed that even if the percent fertility is significantly lower than the parent the 100 grain weight on the average does not differ significantly.

Although the observations made in this particular experiment were in only one season, the results suggest that through irradiation and subsequent anther culture, variabilities may arise which are importing in the development of a variety.

REFERENCES

FOROUGHI-WEHR,B., FRIEDT, W. and WENZEL, G., (1982) On the genetic improvement of androgenetic haploid formation in Hordeum vulgare L. Theor. Appl. Genet. 62: 233-239.

INTERNATIONAL RICE RESEARCH INSTITUTE (1980) Annual report for 1979. Los Banos, Philippines, pp. 538.

NITSCH, C., and GODARD, M., (1979) The role of hormones in promoting and developing growth to select new varieties in sterile culture. In: Plant Regulation and World Agriculture. Plenum Press. NATO Advanced Study on Plant Regulation and World Agriculture, Ismis, 1978, 49-62.

SAMA RAO, H.K. and NARAYANASWAMY, S., (1976) Anatomical anomalies in tissue cultured-induced roots of Cajanus cajan (L.) Mills P. Proc. Indian Acad. Sci. 83B (4): 207-209.

SIGURBJÖRNSSON, B. (1968) Induction of rice breeding with induced mutations. In: Rice Breeding with Induced Mutations. IAEA, Vienna, 3-4.

TORRIZO, L.B. and ZAPATA, F.J., (1986) Anther culture in rice: IV. The effect of abscisic acid on plant regeneration. Plant Cell Reports, 5: 136-139.

ZAPATA, F.J. and ALDEMITA, R.R., (1986) Anther culture of Basmati 370 at IRRI. A. Gamma ray-induced green plant regeneration. IRRN. 11 (4): 22.

Table 1. Anther culture of Basmati 370 in three media.

Media	Callus production (%)	Calli plated (no.)	Calli producing albino plants (no.)	Albino plant production (no.)	(ave.)[a]
E_{10}[a]	5.71	41	16	61	3.8
$E_{10}M$[b]	2.56	10	2	8	4
$E_{10}T$[c]	1.29	6	2	3	1.5

[a]E_{10} consists of Gamborg's B5-basic media with 1 mg 2,4-dichloro-penoxyacetic acid (2,4-D)/1, 0.5 mg 6-benzylaminopurine (BAP)/1, 0.5 mg indole-3-acetic acid (IAA)/1 and 5 g glucose/1.

[b]$E_{10}M$ consists of E_{10} with 0.002 M 2[N-morpholine] ethane sulfonic acid

[c]$E_{10}T$ consists of E_{10} without 2,4-D, but with 1.5 mg 2,4-5 trichlorophenoxyacetic acid/1

Table 2. Agronomic characteristics of anther culture-derived lines from irradiated Basmati 370 (OYT 1986 WS).[1]

Designation	Flowering (days)	Culm length (cm)	Panicle length (cm)	Plant height (cm)	Tiller no.	Fertility (%)	100-grain weight (g)
Basmati 370	92	126	26	152	11	73	1.69
B. 370 (irradiated, 20kr)	86	123	26	149	15	43	1.89
AC5821	88	121	26	147	9	53	1.81
AC5821-2	90	114	25	139	11	30	1.48
AC5821-3	90	115	24	139	9	41	1.66
AC5821-4	92	118	26	144	15	52	1.60
AC5821-5	88	123	24	147	14	42	1.71
AC5821-6	90	130	25	155	10	58	1.61
AC5821-7	88	120	25	145	9	48	1.71
AC5824	90	124	26	150	14	62	1.61
AC5824-3	90	128	25	153	15	62	1.75
AC5824-4	90	128	25	153	9	47	1.65
AC5830	90	128	26	154	12	38	1.34
AC5830-1	90	127	26	153	9	39	1.40
AC5830-2	92	126	27	153	9	49	1.53
AC5835	92	130	25	155	13	63	1.94
AC5835	92	126	26	152	9	49	1.64
AC6166	90	121	26	147	13	54	1.63
AC6166-1	90	123	26	149	11	41	1.60
AC6166-2	88	116	25	141	11	39	1.61
Ave. of AC lines	90	123.2	25.4	148.7	11.2	48.2	1.63

[1] Average of 12 plants per entry.

CROSSABILITY OF SPRING BARLEY MUTANTS WITH H. BULBOSUM

M. GAJ AND M. MALUSZYNSKI*

Department of Genetics
University of Silesia
Katowice, Poland

* Joint FAO/IAEA Division
Plant Breeding Section
Vienna, Austria

M. Maluszynski (ed.), Current Options for Cereal Improvement, 203–210.
© *1989 by Kluwer Academic Publishers.*

ABSTRACT
      Twenty stable mutant lines (M_{10}), their three parents
(Aramir, Georgia and Karat) as well as cultivars Cristal and
Roland were used in studying the heritability of seed set
from pollination with <u>H. bulbosum</u>. 17 Aramir mutants varied
in crossability (seed set percentage) from 2.4% to 80.2%.
Low crossability appears to be dominant in all crosses of
mutants with variety Aramir. Segregation for low and high
crossability observed in the F_2 of two mutants with low
crossability indicated that a minimum of two genes should be
responsible for this character.   Mutagenic treatment of
haploid seeds with MNH was effective. A great number of
morphological changes were observed among doubled haploid
M_2 plants.

INTRODUCTION

The practical application of the "Bulbosum-method" for barley breeding depends mainly on high crossability of both components (Jensen, 1977; Pickering and Hayes, 1976; Simpson et al., 1980). We tried to investigate the inheritance of this character using our large collection of spring barley mutants. Preliminary results of these investigations were presented at the FAO/IAEA Symposium on "Nuclear techniques and in-vitro culture for plant improvement" in 1985 (Gaj and Maluszynski, 1986). During the past year we extended the investigations using additional mutants.

GENETIC ANALYSIS OF CROSSABILITY WITH H. BULBOSUM

Twenty spring barley mutants from our collection were involved in this genetic study. These true breed mutant lines (M_{10}) were obtained after MNH treatment of varieties Aramir, Georgia and Karat. The parent varieties Aramir, Georgia and two other varieties, namely, Cristal and Roland were used, for crosses with H. bulbosum. We used H. bulbosum, diploid form, obtained from T. Adamski of the Institute of Plant Genetics, Poznan, Poland. All plants were grown in a greenhouse at a temperature of 20°C $\pm$ 2°C during the day and 15°C $\pm$ 2°C during the night, with a day length of 16 hours.

The crossability of different genotypes was defined as the number of seeds obtained after pollination by H. bulbosum. We found great differences among genotypes in seed set after pollination. Varieties Aramir and Roland produced the highest number of seeds, namely 86.5% and 69.4% of pollinated flowers, respectively. Contrary to this, varieties Georgia and Cristal had the lowest number, 3.8% and 1.5%. 17 Aramir mutants varied in seed set percentage - from 80.2% to 2.4%.

We investigated only two mutants from the variety Georgia, which had the same poor crossability as the parent variety. The large variation observed in this material with regard to crossability with H. bulbosum gave us the opportunity to study its inheritance.

Mutants or varieties with different crossability to H. bulbosum were used as cross components. The variety Aramir was the other parent in each cross. All F_1 plants from crosses variety Aramir x mutants with very low crossability, produced very few seeds after pollination by H. bulbosum. When two varieties (Georgia and Cristal) were used as the second parent in these crosses, we had the same outcome. This indicated that the low crossability character is determined, in investigated material, by dominant factor(s). Five such mutants were obtained after mutagenic treatment of the variety Aramir, which in our experiment expressed the highest crossability with H. bulbosum. It seems that the low crossability character can very easily be induced by mutagen treatment.

The dominant nature of low crossability with H. bulbosum was found as well in crosses with two botanical varieties of wild barley: H. spontaneum var. bactrianum Vav. and H. spontaneum var. transcapicum Vav.

To confirm the presented thesis we analyzed the F_2 progeny of the cross Aramir with its mutant 104 AR which expressed low crossability (11,8%). Individual crossing of each F_2 plant with H. bulbosum, demonstrated that from 198 investigated plants, 152 expressed low crossability but 46 behaved similar to Aramir in this character (Fig. 1). The seed set in the first group ranged from 0-25%, but most of these plants (90) were found in the 0-5% class of the seed set. The seed set in the group of recessive genotypes was 55-100%. A segregation ratio of 3:1 (low to high crossability) was found with X^2 = 0.28. The results suggest that a monogenic mutation is responsible for low crossability of the mutant 104 AR .

Mutant 104 AR has a six-row spike, determined by gene v located on barley chromosome 2 and black lemma with pericarp, which is determined by gene B on chromosome 5 (Gaj and Maluszynski, 1985). The analysis of recombinants in the above-mentioned F_2 progeny did not indicate any linkage between these genes and the locus responsible for crossability with H. bulbosum.

An interesting result was obtained from the pollination of a two mutants hybrid with very low crossability (322 GR x 099 AR) with H. bulbosum. F_1 plants from this cross, pollinated by H. bulbosum, produced a low seed set (12.2%). 160 plants of F_2 progeny behaved in a similar way, while 46 other F_2 plants from the same progeny produced seeds at a much higher frequency and 15 plants had a crossability similar to the variety Aramir (Fig. 2). The appearance of forms with high crossability in progenies of two mutants with very low crossability indicated that at least two genes should be responsible for this character, though the belief until now has been that only one gene, located on chromosome 7, was responsible for crossability (Pickering, 1983).

NEW INDUCED MUTATION EXPERIMENTS

As a result of the final FAO/IAEA Research Co-ordination Meeting on "Semi-dwarf mutants for cereal improvement" which was held in Rome 1985, we began experiments using mutation induction for the purpose of increasing the variability observed among doubled haploids of cereals. The use of mutagens in connection with the doubled haploid methods gives the opportunity to produce haploid seedlings with mutations in the hemizygotic stage. The duplication of the number of chromosomes in such material results in a pure, non-chimeric, homozygous mutant line, thus accelerating the mutation breeding procedure by at least 5-6 generations.

We can present only preliminary results of experiments using three mutation induction methods to increase the variability of doubled haploids obtained by the "Bulbosum method" from variety Aramir. Insufficient material made it impossible to optimize the dose for the mutagens MNH and EI. For this reason the two methods tried, namely injection of mutagen to leaf sheath below the 2 cm long spike or treatment of haploid seedlings were very ineffective and resulted in the low rate of survival and low fertility of DH plants. The third method (treatment of haploid seeds by mutagen, MNH) gave good results. From the application of 1 mM concentration of MNH for 1 or 2 hours, 127 haploid M_1 plants were obtained. After duplication of chromosomes, 41 fertile DH M_1 plants were grown to maturity and produced M_2 seeds. In total, from 1600 treated haploid seeds we obtained 370 M_2 plants, with a great number of changes in different characters such as plant height, tillering, time of maturity, etc. The genetic stability of these DH mutants is now under investigation. The crossability of these mutants has not been yet studied.

REFERENCES:

GAJ, M. and MALUSZYNSKI, M., (1985) Genetic analysis of spike characters of barley mutants. Barley Genetics Newsletter 15: 32-33

GAJ, M. and MALUSZYNSKI, M., (1986) Genetic studies of crossability of barley mutants with Hordeum bulbosum L. and Secale cereale L. In: Nuclear Techniques and In Vitro Culture for Plant Improvement, IAEA, Vienna: 221-226

JENSEN, C.J., (1977) Monoploid production by chromosome elimination. In: Applied and Fundamental Aspects of Plant Cell, Tissue and Organ Culture, Reinert, J. and Y.P.S. Bajaj, Springer-Verlag, Berlin: 299-340

PICKERING, R.A. (1983) The location of a gene for incompatibility between Hordeum vulgare L. and H. bulbosum L. Heredity 59: 455-459

PICKERING, R.A. and HAYES, J.D., (1976) Partial incompatibility in crosses between Hordeum vulgare L. and H. bulbosum L. Euphytica 25: 671-678

SIMPSON, E., SNAPE, J.W. and FINCH, R.A., (1980) Variation between Hordeum bulbosum genotypes in their ability to produce haploids of barley, Hordeum vulgare. Z. Pflanzenzüchtg. 85: 205-211

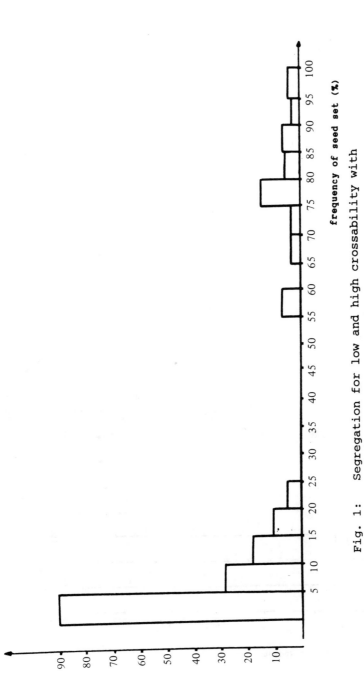

Fig. 1: Segregation for low and high crossability with
H. bulbosum L. in F_2 of mutant 104 AR x original
variety Aramir.

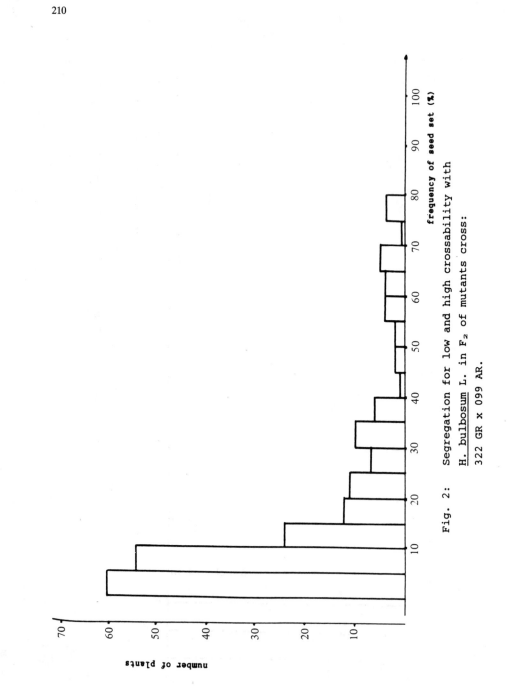

Fig. 2: Segregation for low and high crossability with
         H. bulbosum L. in F_2 of mutants cross:
         322 GR x 099 AR.

THE COMMERCIAL APPLICATION OF CEREAL HAPLOIDY

A. MARSOLAIS

Kyasma Gene Resources Ltd.
R.R. #2, Alma, Ontario
Canada

M. Maluszynski (ed.), Current Options for Cereal Improvement, 211–214.
© *1989 by Kluwer Academic Publishers.*

Abstract

Kyasma Gene Resources Ltd. was incorporated to commercialize wheat doubled haploid production technology. This paper describes Kyasma's production sequence and research goals.

Commercialization of wheat doubled haploids

Kyasma Gene Resources Ltd. (Kyasma) was incorporated in 1986 to commercialize the wheat doubled haploid (DH) production technology developed at the University of Guelph under the leadership of Professor K.J. Kasha. Kyasma offers a wheat DH production service to wheat breeders in North America. We produce homozygous DH lines from hybrid or heterozygous parental lines for a price of $16 Canadian per DH line. It takes approximately 12-14 months to produce DH lines from winter wheat parents and 9-11 months from spring wheat parents. In our production sequence, we grow the parent plants, culture the anthers at the proper stage, transfer the embryoids to a regeneration medium, regenerate plants, double the chromosomes using colchicine, grow out the DH plants, then ship the mature DH seeds back to the customer. It is anticipated that in the 1986/87 production year approximately 4500 DH's will be produced. The composition of the basic culture media and related techniques have been published (Marsolais, 1986; Marsolais and Kasha, 1985).

The research at Kyasma over the past year has concentrated on improvements in the rate of plant regeneration and the substitution of lower cost medium components. Regeneration from wheat microspore-derived embryoids is now highly efficient with greater than 90% of the embryoids producing plants. Albinism is still a problem in some genotypes. The frequency of albinism has ranged from approximately 20% to 50% with a mean of approximately 30% of the regenerated plants. The composition of the regeneration medium has not influenced the proportion of albino plants produced.

For several months we have been working on the technique of isolated microspore culture and recently have been able to induce large numbers of microspores through the first few cell divisions in vitro. However, after 7-10 days in culture, the mortality rate has escalated rapidly. The isolated microspore culture technique at present is not as efficient or reliable for embryoid production as the anther culture technique. We will continue working on isolated microspore culture this year.

An efficient microspore culture technique is the first step in developing a new approach to cereal breeding. Further research will deal with in vitro chromosome doubling, where the goal is to produce DH embryoids directly from microspores. This will be followed by embryo dry-down or encapsulation that will allow direct seeding of DH embryos in the field. Plant breeders will be able to produce pure lines (DH's) in the same time it takes to produce the first segregating generation (usually called the F_2 generation when produced by self-pollinating an F_1 hybrid). There will also be a considerable cost saving relative to conventional procedures for pure line production. We expect

214

that the research leading to the development of this system will take several years; however we feel that the economic and social benefits will more than offset the time and money expended.

We also are engaged in research with spring durum wheat and spring barley. Albinism and poor rates of plant regeneration from calli remain the major stumbling blocks to the use of barley anther culture in barley breeding. We have not made any progress in solving either of these problems. We currently are testing our standard wheat DH production system on a range of durum cultivars and hybrids provided by Dr. D.R. Knott of the University of Saskatchewan.

Will commercial DH production be a viable, profitable business? At this time, we don't have a firm answer. Our unit production cost is close to the $16 per DH line that we are charging. Private corporations and public institutions involved in crop research are under considerable stress due to the lengthening depression in the agriculture sector. If our sales volume in the near future does not justify the capital and labour expended, we will not offer this service. However, we will continue to use DH technology in our own spring wheat breeding programme.

References

MARSOLAIS, A.A. (1986) Callus induction from barley microspores. Ph.D Thesis, University of Guelph, Guelph.

MARSOLAIS, A.A., KASHA, K.J. (1985) Callus induction from barley microspores. The role of sucrose and auxin in a barley anther culture medium. Can. J. Bot. 63: 2209.